T0191703

Big-Data Analytics and Cloud Computing

Marcello Trovati • Richard Hill • Ashiq Anjum
Shao Ying Zhu • Lu Liu
Editors

Big-Data Analytics and Cloud Computing

Theory, Algorithms and Applications

 Springer

Editors
Marcello Trovati
Department of Computing
   and Mathematics
University of Derby
Derby, UK

Ashiq Anjum
Department of Computing
   and Mathematics
University of Derby
Derby, UK

Lu Liu
Department of Computing
   and Mathematics
University of Derby
Derby, UK

Richard Hill
Department of Computing
   and Mathematics
University of Derby
Derby, UK

Shao Ying Zhu
Department of Computing
   and Mathematics
University of Derby
Derby, UK

ISBN 978-3-319-79767-0   ISBN 978-3-319-25313-8   (eBook)
DOI 10.1007/978-3-319-25313-8

Springer Cham Heidelberg New York Dordrecht London
© Springer International Publishing Switzerland 2015
Softcover reprint of the hardcover 1st edition 2015

Printed on acid-free paper

Springer International Publishing AG Switzerland is part of Springer Science+Business Media (www. springer.com)

Foreword

Among developments that have led to the domain of cloud computing, we may consider the following. Very often, the workplace is now distributed and potentially even global. Next, there is the ever-increasing use being made of background 'big data'. When data is produced in real time and dynamically evolving, then a cloud platform is highly beneficial. Next comes the wide range of platforms used for access and use of data and information. In this picture, mobile and networked platforms are prominent. So too are the varied aspects of pervasive and ubiquitous computing and systems.

Cloud platforms are the foundations for our physical and virtual environments that are empowered increasingly by the Internet of Things. That is, the general progression that enables smarter cities and other related developments. Among these also are the smart workplace and the smart learning environment.

This book collects together many discussion and research topics relating to cloud services, technologies and deployments. Included are cloud service provision, integration with advanced interactivity and cloud-based architectures for the provision of large-scale analytics. Sustainability plays a crucial role, especially in relation to data centres, data grids and other layers of middleware that can be central parts of our compute environment and data clouds. The following inspirational quotation was voiced by Christian Belady, General Manager, Data Center Services, Microsoft: 'Data is really the next form of energy ... I view data as just a more processed form of energy'.

The contributions in this book aim at keeping one fully abreast of these big data and closely related developments. Even more rewarding is to be actively engaged in such technological progress. It can well be the case that dividing lines effectively disappear in regard to user and supplier and producer and consumer, where the latter becomes the prosumer.

The reader can enjoy this book's contents, and draw inspiration and benefit, in order to be part of these exciting developments.

Big Data Laboratory   Professor Fionn Murtagh
University of Derby, UK
August 2015

Preface

Overview and Goals

Data is being created around us at an increased rate, in a multitude of forms and types. Most of the advances in all the scientific disciplines that have occurred over the last decade have been based on the extraction, management and assessment of information to provide cutting-edge intelligence. This, in turn, has accelerated the need, as well as the production of large amounts of data, otherwise referred to as big data.

Due to the diverse nature of big data, there is a constant need to develop, test and apply theoretical concepts, techniques and tools, to successfully combine multidisciplinary approaches to address such a challenge. As such, theory is continuously evolving to provide the necessary tools to enable the extraction of relevant and accurate information, to facilitate a fuller management and assessment of big data.

As a consequence, the current academic, R&D and professional environments require an ability to access the latest algorithms and theoretical advance in big data science, to enable the utilisation of the most appropriate approaches to address challenges in this field.

Big-Data Analytics and Cloud Computing: Theory, Algorithms and Applications presents a series of leading edge articles that discuss and explore theoretical concepts, principles, tools, techniques and deployment models in the context of Big Data.

Key objectives for this book include:

- Capturing the state of the art in architectural approaches to the provision of cloud-based big data analytics functions
- Identifying potential research directions and technologies to facilitate the realisation of emerging business models through big data approaches
- Providing relevant theoretical frameworks and the latest empirical research findings

- Discussing real-world applications of algorithms and techniques to address the challenges of big data-sets
- Advancing understanding of the field of big data within cloud environments

Organisation and Features

This book is organised into two parts:

- Part I refers to the theoretical aspects of big data, predictive analytics and cloud-based architectures.
- Part II discusses applications and implementations that utilise big data in cloud architectures.

Target Audiences

We have written this book to support a number of potential audiences. *Enterprise architects* and *business analysts* will both have a need to understand how big data can impact upon their work, by considering the potential benefits and constraints made possible by adopting architectures that can support the analysis of massive volumes of data.

Similarly, *business leaders* and *IT infrastructure managers* will have a desire to appreciate where cloud computing can facilitate the opportunities afforded by big data analytics, both in terms of realising previously hidden insight and assisting critical decision-making with regard to infrastructure.

Those involved in system design and implementation as *application developers* will observe how the adoption of architectures that support cloud computing can positively affect the means by which customers are satisfied through the application of big data analytics.

Finally, as a collection of the latest theoretical, practical and evaluative work in the field of big data analytics, we anticipate that this book will be of direct interest to *researchers* and also *university instructors* for adoption as a course textbook.

Suggested Uses

Big-Data Analytics and Cloud Computing can be used as an introduction to the topic of big data within cloud environments, and as such the reader is advised to consult Part I for a thorough overview of the fundamental concepts and relevant theories.

Part II illustrates by way of application case studies, real-world implementations of scenarios that utilise big data to provide value.

Readers can use the book as a 'primer' if they have no prior knowledge and then consult individual chapters at will as a reference text. Alternatively, for *university instructors*, we suggest the following programme of study for a twelve-week semester format:

- Week 1: Introduction
- Weeks 2–5: Part I
- Weeks 5–11: Part II
- Week 12: Assessment

Instructors are encouraged to make use of the various case studies within the book to provide the starting point for seminar or tutorial discussions and as a means of summatively assessing learners at the end of the course.

Derby, UK   Marcello Trovati
                                                                                       Richard Hill
                                                                                       Ashiq Anjum
                                                                                     Shao Ying Zhu
                                                                                            Lu Liu

Acknowledgements

The editors acknowledge the efforts of the authors of the individual chapters, without whose work, this book would not have been possible.

Big Data Laboratory
Department of Computing and Mathematics
University of Derby, UK
August 2015

Marcello Trovati
Richard Hill
Ashiq Anjum
Shao Ying Zhu
Lu Liu

Contents

Contributors

Ahmad Faisal Abidin Computing Science and Mathematics, University of Stirling, Stirling, Scotland, UK

Ashiq Anjum Department of Computing and Mathematics, University of Derby, Derby, UK

Omar Behadada Department of Biomedical Engineering, Faculty of Technology, Biomedical Engineering Laboratory, University of Tlemcen, Tlemcen, Algeria

Nik Bessis Department of Computing and Mathematics, University of Derby, Derby, UK

Lewis Craske Department of Computing and Mathematics, University of Derby, Derby, UK

Dominic Davies-Tagg Department of Computing and Mathematics, University of Derby, Derby, UK

Kurt Englmeier Faculty of Computer Science, Schmalkalden University of Applied Science, Schmalkalden, Germany

Jer Hayes IBM Research, Dublin, Ireland

Richard Hill Department of Computing and Mathematics, University of Derby, Derby, UK

Paul Holmes Department of Computing and Mathematics, University of Derby, Derby, UK

Amir Hussain Computing Science and Mathematics, University of Stirling, Stirling, Scotland, UK

Ferosh Jacob Data Science R&D, CareerBuilder, Norcross, GA, USA

Faizan Javed Data Science R&D, CareerBuilder, Norcross, GA, USA

Aaron Johnson Department of Computing and Mathematics, University of Derby, Derby, UK

Mario Kolberg Computing Science and Mathematics, University of Stirling, Stirling, Scotland, UK

Peter Larcombe Department of Computing and Mathematics, University of Derby, Derby, UK

Demetrio Mestre Department of Computer Science, Federal University of Campina Grande, Campina Grande, PB, Brazil

Dimas C. Nascimento Department of Computer Science, Federal University of Campina Grande, Campina Grande, PB, Brazil

Carlos Eduardo Pires Department of Computer Science, Federal University of Campina Grande, Campina Grande, PB, Brazil

Marcello Trovati Department of Computing and Mathematics, University of Derby, Derby, UK

Part I
Theory

Chapter 1
Data Quality Monitoring of Cloud Databases Based on Data Quality SLAs

Dimas C. Nascimento, Carlos Eduardo Pires, and Demetrio Mestre

Abstract This chapter provides an overview of the tasks related to the continuous process of monitoring the quality of cloud databases as their content is modified over time. In the Software as a Service context, this process must be guided by data quality service level agreements, which aim to specify customers' requirements regarding the process of data quality monitoring. In practice, factors such as the Big Data scale, lack of data structure, strict service level agreement requirements, and the velocity of the changes over the data imply many challenges for an effective accomplishment of this process. In this context, we present a high-level architecture of a cloud service, which employs cloud computing capabilities in order to tackle these challenges, as well as the technical and research problems that may be further explored to allow an effective deployment of the presented service.

1.1 Introduction and Summary

Data quality monitoring (DQM) is the process that evaluates a data set to determine if it meets the planning objectives of a project and thus provide data of the right type, quality, and quantity to support their intended use. DQM is built on a fundamental premise: data quality is meaningful only when it relates to the intended use of the data. For example, in order to achieve the full potential that is provided by an integrated view of the data integrated in a data warehouse, it is necessary to develop strategies to maintain acceptable levels of confidence over the data. Similarly, other levels of confidence must also be consolidated regarding legal aspects and industry standards. Thereby, establishing data quality standards and adopting a continuous monitoring strategy in order to guarantee that these standards are used will lead to an overall data quality improvement. In turn, this accomplishment will result in reduction of time spent on diagnosis and correction, increased speed of delivering information, and improvement of confidence in the decisions [1]. In this context,

D.C. Nascimento (✉) • C.E. Pires • D. Mestre
Department of Computer Science, Federal University of Campina Grande, Rua Aprigio Veloso, 882, 58429-900 Campina Grande, PB, Brazil
e-mail: dimascnf@gmail.com

© Springer International Publishing Switzerland 2015
M. Trovati et al. (eds.), *Big-Data Analytics and Cloud Computing*,
DOI 10.1007/978-3-319-25313-8_1

there is an evident need to incorporate data quality considerations into the whole data cycle, encompassing managerial/governance as well as technical aspects [2].

Regarding the Big Data initiative, data governance must ensure high data quality as a basis for its effective usage [3]. A Big Data user may focus on quality which means not having all the data available but having a (very) large quantity of high-quality data that can be used to draw precise and high-valued conclusions [4]. In fact, one of the focuses of Big Data systems should be on quality data storage rather than storing very large irrelevant data. Thereby, it is important to investigate questions such as deciding which portion of data is relevant, how much data would be enough for decision making, and whether the stored data is accurate (or not) to draw conclusions from its content [5]. All these questions are strongly tied to data quality concepts and dimensions.

In this sense, the process of DQM is challenged by many factors such as (i) the amount of processed data (Big Data era), (ii) the heterogeneity of data sources and structure, (iii) the computational complexity of the data quality algorithms, (iv) the amount of hardware infrastructure required to execute these algorithms, and (v) the velocity of changes (insertions, updates, and deletions) that affects the data stored in databases. Due to these challenges, business managers may prefer to outsource the overall process of data storage and continuous monitoring of data quality, due to either operational or financial reasons. Nowadays, hardware- and service-level outsourcing is usually done by using cloud computing technologies. Cloud computing has recently emerged as a computing platform with reliability, ubiquity, and availability in focus [6] mainly by utilizing computing as an on-demand service [7] and simplifying the time-consuming processes of hardware provisioning, hardware purchasing, and software deployment. Another advantage is scalability in terms of computing resources. Service providers can scale up when the demand for service rises significantly. Similarly, they can scale down when the demand decreases. Cloud computing also enables service consumers to use services on a pay-as-you-go or a monthly subscription basis [6].

Regarding the data quality monitoring outsourcing contract, the rules and requirements that need to be met by the service provider are described in a data quality SLA (DQSLA), which is the most common way to specify contract parameters in the cloud computing context. In this scenario, this book chapter has the following main objectives:

- *The DQSLA formalization*: this formalism will allow the representation of unambiguous agreements between customers and data quality service providers.
- *A data quality-aware service architecture*: a high-level architectural design that depicts the main modules of a service to provide the process of continuous data quality monitoring, based on DQSLA inputs and suitable approaches for handling the Big Data scale.
- *Open research problems*: the discussion of a catalog of open research problems that may be further explored by database researchers and practitioners.

These objectives are further detailed and discussed in the following sections of this chapter.

1.2   Background

Any data quality monitoring life cycle will include an evaluation (assessment) phase, which intends to access and process the data according to data quality objectives (that in turn, need to take into account the business objectives). In this step, each data quality objective may generate many data quality rules, which are mapped to data quality algorithms. Depending on the complexity of these algorithms and on the amount of data that need to be processed, the process of data quality monitoring may require a large amount of computational resources. For example, a record linkage task (which aims to identify duplicated records stored in the database) presents high computational complexity. When two database tables, A and B, are to be matched to identify possible duplicates, potentially each record from A needs to be compared with every record from B, resulting in a maximum number of $(|A| \times |B|)$ comparisons between two records [8]. Therefore, the computational efforts of comparing records increase quadratically as databases get larger. Moreover, the data quality evaluation needs to be frequently re-executed when the data is updated over time.

1.2.1   Data Quality in the Context of Big Data

Nowadays, data is growing at a huge speed, making it difficult to handle such large amount of data (exabytes) [5]. Similar to what happens to traditional systems, classical algorithms are not designed to handle the Big Data scale and demands. Big Data imposes two initial classes of challenges: engineering, efficiently managing data at unimaginable scale, and semantics, finding and meaningfully combining information that is relevant to its intended usage [9]. In turn, when handling the semantics challenge, its accomplishment may be strongly influenced by the quality of the data, and thus data quality evaluation, together with efficient management of Big Data, becomes two essential concerns. For example, it is important to measure completeness and consistency, which are data quality dimensions, of Big Data data sets [9]. Clearly, effective Big Data processing involves a multidisciplinary approach toward its solution.

In order to accomplish a meaningful usage of Big Data, it is necessary to perform an effective usage of computation capacity and algorithms to integrate, filter, analyze, process, and identify patterns using Big Data data sets and, based on the results, draw useful conclusions and claims. This scenario of Big Data data sets claims a wise usage and employment of current technologies for an effective usage of this opportunity. The employment of suitable technologies and approaches toward Big Data will allow practitioners and researchers to ask crucial questions about the meaning of the data, evaluate how trustable is its content, and decide in which contexts to apply the discoveries, in other words, how to get the true value of Big Data.

In practice, the three main V's (velocity, variety, volume) of a Big Data application impose many challenges for the tasks of data quality evaluation, monitoring, and correction. First, the Big Data scale may generate very high execution times for an evaluation of its quality when employing naive approaches. In turn, the variety of data types and formats that are present in Big Data applications demands new breakthroughs on the data quality algorithms. Lastly, the velocity of data updates is often high, quickly making previous data quality results obsolete [10]. In this context, Big Data's success is inevitably linked to an intelligent management of data selection and usage as well as joint efforts toward clear rules regarding data quality [3]. To this end, tasks such as filtering, cleansing, pruning, conforming, matching, joining, and diagnosing should be applied at the earliest touch points possible [5].

Therefore, it becomes important to design new classes of systems and algorithms that are able to scale up to thousands or even millions of entities [11]. This can be accomplished by combining efficient approaches for executing data quality algorithms in a distributed manner [12, 13] together with optimized techniques for executing data quality algorithms, such as indexing [14] and incremental approaches [10]. These approaches are further discussed and detailed throughout this chapter.

1.2.2 Cloud Computing

Cloud services are applications or services offered by means of cloud computing. Therefore, by adopting cloud services, business managers are considering to take advantage of the economic benefits offered by maintaining parts of its IT resources and tasks by a cloud service provider [6]. Nowadays, nearly all large software companies, such as Google, Microsoft, and Oracle, are providing cloud services. Besides, cloud computing has revolutionized the standard model of service provisioning, allowing delivery over the Internet of virtualized services that can scale up and down in terms of processing power and storage. Cloud computing also provides strong storage, computation, and distributed capability to support Big Data processing. In order to achieve the full potential of Big Data, it is required to adopt both new data analysis algorithms and new approaches to handle the dramatic data growth and needs of massive scale analytics. As a result, one of the underlying advantages of deploying services on the cloud is the economy of scale. By using the cloud infrastructure, a service provider can offer better, cheaper, and more reliable services [6].

The high-level architecture of the cloud computing business model is shown in Fig. 1.1. The architecture depicts four levels of the stack: client layer, service layer, platform layer, and infrastructure layer. Service providers deploy their services (Software as a Service level (SaaS)) on a cloud computing environment that can be used or consumed by customers or other applications on the client layer. The consumption of the available services is formalized by a service level agreement (SLA) that is arranged between the customer and the provider of the service. In

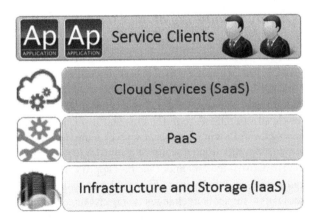

Fig. 1.1 High-level cloud computing stack

the SaaS level, the customers do not have control over the hardware- and software-level configurations of the consumed service. In practice, the interaction between the customers and the service is limited to the service interface and customers' input parameters. The data quality-aware service architecture, further discussed in this book chapter, is an example of a SaaS that needs to meet predefined quality of service requirements (specified as data quality SLAs).

Alternatively, customers may also hire the second level of the stack (Platform as a Service (PaaS)), which is a development platform that facilitates the development of cloud services by the customers. This platform usually includes frameworks, developing and testing tools, configuration management, and abstraction of hardware-level resources. Lastly, customers can hire hardware-level resources available in the Infrastructure as a Service (IaaS) cloud. Virtualization is extensively used in IaaS cloud in order to integrate/decompose physical resources in an ad hoc manner to meet growing or shrinking resource demand from cloud consumers [15]. Regardless of the level of the cloud stack that is consumed by the customers, the consumed resources (either physical or virtual) are usually delivered using a predefined pay-per-use model.

A cloud database is a database that is hosted on a cloud computing environment. In practice, the access to a cloud database can be performed using a database management system (DBMS) that runs over a virtual machine or via a database service interface. In the latter case, usually named Database as a Service (DBaaS), the database installation, maintenance, and accessibility interface are provided by a database service provider. The adoption of DBaaS is an effective approach to cloud-based applications and services as part of the Software as a Service business model. Among the main characteristics of a DBaaS, there are cloud portability, scalability, and high availability. These characteristics are provided by using cloud computing capabilities, such as hardware provisioning and service redundancy.

1.2.3   Data Quality Monitoring in the Cloud

A possible approach to perform data quality monitoring in the cloud is by adopting a data quality-aware service (DQaS) [16]. This service relies on data quality service level agreements (DQSLAs) established between service consumers and the DQaS provider.

In order to tackle strict DQSLA requirements and Big Data processing, a DQaS must use or combine the following approaches: (i) execute incremental data quality algorithms, (ii) use computational resources provisioning, or (iii) execute data quality algorithms in a distributed manner. The first approach has been addressed by the scientific community, mainly by using historical metadata to avoid reprocessing an entire data set D after subsequent changes ($\triangle D_1$, $\triangle D_2$, ... , $\triangle D_n$) over D [10]. For doing so, metadata about the initial data set evaluation is stored, which is then used to avoid unnecessary comparisons and computations over the portion of D that is not related or will not be influenced by a specific $\triangle D_i$ that affected D.

The main goal of this chapter is to discuss the adoption of the approaches (ii) and (iii) on a DQaS architecture that relies on DQSLAs. In other words, we intend to explain how to take advantage of cloud computing capabilities in order to tackle customers' DQSLA requirements and Big Data processing.

1.2.4   The Challenge of Specifying a DQSLA

In the context of cloud computing, a quite relevant metric is the level of satisfaction of cloud customers in terms of performance and quality of service (QoS) they get from cloud service providers. QoS refers to a set of qualities or characteristics of a service, such as availability, security, response time, throughput, latency, reliability, and reputation. Such qualities are of interest for service providers and service consumers alike [6]. The agreement between the customer and the service provider, known as the service level agreement (SLA), describes agreed service functionality, cost, and qualities [17].

In essence, an SLA is a legally binding contract that states the QoS guarantees that an execution environment, such as a cloud-based platform, has to provide its hosted applications with [18]. An SLA consists of a collection of contractual clauses between a service provider and a service customer that detail the nature, quality, and scope of the service to be provided and its authorized use [19]. Finally, an SLA may also specify the percentage of SLA violations that be tolerated, within a predefined time interval, before the service provider incurs a (e.g., economic) penalty [18].

Measuring conformance to defined service levels has long been part of network, server, and desktop management [1]. For example, it is usually clear when the network is too slow or if a server fails. However, the difference between hardware service level management and data quality service level management is in the perceived variability in specifying what is meant by *acceptable levels of service* [1] and how to represent these *levels of data quality acceptability*. In other words, in practice it is a challenge to represent a data quality service level agreement (DQSLA). Therefore, we state the necessity of using a formal structure that is dynamic and flexible enough to express service customer's QoS requirements as well as service providers' capabilities and offers [16], both in the data quality context. In other words, we need to represent high-level business objectives regarding data quality as low-level thresholds for data quality algorithms. Thereby, by listing the critical expectations, methods for measurement, and specific thresholds, the business clients can associate data governance with levels of success in their business activities [1].

1.2.5   The Infrastructure Estimation Problem

One of the approaches that can be used by a DQaS to tackle DQSLA requirements and Big Data handling, by means of cloud computing, is to use infrastructure provisioning, i.e., an on-demand allocation of hardware provisioning.

In order to execute a data quality algorithm to evaluate database data, a DQaS must allocate a computational grid to execute the algorithm in a distributed manner. In practice, the service's resource requirements are fulfilled by the physical or virtual computational nodes (virtual machines) upon which the algorithms are executed [16]. The configuration class estimated by the service is a cluster composed by a pair $<$ Number of Nodes (VM's), VM Configuration $>$. Let $N = \{N_1, N_2, \ldots, N_n\}$ be the amount of VM's that can be allocated for an execution of a data quality algorithm and $\gamma = \{\gamma_1, \gamma_2, \ldots, \gamma_k\}$ the available VM configurations. A configuration class (Cl) is a pair $< N_i, \gamma_j >$ that represents the configuration of a grid composed by virtual machines. Then, $Cl = (N \times \gamma) = \{Cl_1, Cl_2, \ldots, Cl_{n.k}\}$ represents the set of all possible configuration classes that can be chosen by the DQaS.

Intuitively, one of the critical parts of a DQSLA is the time restriction to execute a data quality task. This is particularly important since the data quality monitoring process may affect how fast users and systems may rely on the monitored data sets to use the data for their respective purposes. In this context, for the execution of a single data quality algorithm execution e, we want to estimate a proper class Cl_c that is able to minimize the difference between the restriction time defined on the DQSLA that triggered e (Tres(e)) and the execution time of e (ExecTime(e)), as stated in Table 1.1.

If an underprovisioning policy is adopted, i.e., allocating a low configuration class for the execution of a data quality algorithm, it is highly probable that the

Table 1.1 The infrastructure estimation problem

Find	Cl_c
Over	$(N \times \gamma) = \{Cl_1, Cl_2, \ldots, Cl_{n.k}\}$
Subject to	ExecTime(e) $\leq$ Tres(e)

DQSLA time restriction will not be met by the DQaS. On the other hand, in order to ensure that a customer DQSLA time restriction is not violated, a resource overprovision policy may be adopted, which is based on evaluating (either through application modeling or through application benchmarking) all possible resources that a data quality algorithm can require in the worst case and then statically allocating these resources to that particular execution. This policy can lead to a largely suboptimal utilization of the cloud environment resources. In fact, being based on a worst-case scenario, a number of allocated resources may remain unused at run time. This limitation can be overcome by developing provisioning algorithms, which can be integrated in the DQaS architecture in order to manage dynamically the cloud configuration classes for the data quality algorithms' executions and honor the DQSLA time restrictions using a minimum amount of virtualized infrastructure. This approach is further detailed in the following sections of this chapter.

1.3 Proposed Solutions

In this section, we discuss potential design solutions to the problems described throughout this book chapter, namely, (i) the DQSLA representation, (ii) the infrastructure provisioning problem, and (iii) the challenges that need to be tackled regarding the DQaS development.

1.3.1 Data Quality SLA Formalization

In order to allow an unambiguous and formal representation of a data quality SLA, in this section we formalize its structure, by enabling the involved parties to express their expectations regarding information such as the data quality dimensions, parameters of the data quality algorithms, the data sets to be analyzed, the overall efficiency of the process, and so on. We define a data quality SLA [16] as a 9-tuple:

$$\textbf{DQSLA} = <R_{id}, DQ_{dim}, M_{rules}, \Delta ts_{valid}, |\Delta D|_{thr}, T_{res}, R_{method}, SLA_{penalties},$$

$$Init_{exec} >, \text{ where:}$$

1. $\textbf{R}_{id}$: is a unique resource identifier of the data set (or data sets involved) to be monitored
2. $\textbf{DQ}_{dim}$: is a data quality dimension [20], such as completeness, accuracy, duplication, volatility, and so on
3. $\textbf{M}_{rules}$: represents the details (parameters of the data algorithms) used to evaluate the monitored data set (referenced as R_{id})
4. $\Delta \textbf{ts}_{valid}$: is the timestamp interval in which the DQSLA is valid
5. $|\Delta \textbf{D}|_{thr}$: is the amount of changes (insertion, update, or deletion) in the data set (R_{id}) that will trigger a new execution of a data quality algorithm by the service to evaluate the quality of the data set as specified by the M_{rules} parameter

6. T_{res}: is the expected efficiency of a single evaluation (execution of a data quality algorithm) of the data set R_{id}
7. R_{method}: is the method (Online or Historical) of reporting the subsequent results of the evaluations of the data set R_{id}
8. $SLA_{penalties}$: defines the penalties for the cases that the defined restriction time (T_{res}) is not met by the service
9. $Init_{exec}$: is a boolean flag that indicates if the data set (R_{id}) must be initially evaluated

1.3.2 Examples of Data Quality SLAs

The process of data quality assessment is performed based on data quality dimensions to be evaluated. In turn, a data quality dimension needs to be further detailed by specifying specific parameters for data quality algorithms and thresholds that define the levels of acceptability for a specific aspect of the data. For this reason, some of the inputs of the DQSLA may be specified based on predefined Backus Normal Form (BNF) rules, for example:

$[R_{method}]$:= Online | Historical
$[DQ_{dim}]$:= Accuracy | Completeness | Timeliness | Duplication | . . .

Using the above rules, we provide the following example of a DQSLA using the Duplication [21] dimension:

DQSLA_01 = < 011, Duplication, $Dedup_{rules}$, (2014-11-09 11:20, 2014-11-11 11:20), 5000, 15 min, Online, $SLA_{penalties}$, true >

In practice, the DQSLA_01 SLA specifies that the data set (011) must be continuously monitored for duplication detection during the timestamp interval (2014-11-09 11:20, 2014-11-11 11:20). The evaluation must be performed according to the parameters specified by the $Dedup_{rules}$ content and must not take more than 15 min to be executed. Besides the initial evaluation of the data set ($Init_{exec}$ = true), the data set (011) must be reevaluated after each 5000 changes on its content. Moreover, the results of the evaluation of the monitored data set should be continuously reported via a web interface (R_{method} = Online). Lastly, there is a penalty for the service provider ($SLA_{penalties}$) if the evaluation process takes more time than 15 min.

In turn, the $Dedup_{rules}$ component of the DQSLA may be further detailed according to the following BNF rules:

```
Dedup_rules = {[Detection_Method], [Similarity_Function],
    [Match_Rules], [DQ_Dim_Evaluation]}
[Detection_Method] := Standard_Blocking |
    Graph_Clustering | Sorted_Neighborhood | ...
```

```
[Similarity_Function] := Jaccard | Jaro_Winkler |
  Edit_Distance | n-gram | ...
[Match_Rules] := ([Sim_Value] [θ] [Thr_Value]
  => [Sim_Class]) |
                     ([Sim_Value] [θ] [Thr_Value]
  => [Sim_Class]  AND [Match_Rules])
[θ] := > | < | ≥ | ≤ | ≠ | =
[Sim_Class] := Match | Non_Match | Possible_Match
[Dim_Evaluation] := (|matches| / |resource_id|)
[Sim_Value] := [d] . [d]
[Thr_Value] := [d] . [d]
[blocking_key] := [attr_name] | substring([attr_name],
  [d], [d]) | [blocking_key] concat [blocking_key]
[attr_name] := [letter] [character]*
[character]:= [d] | [letter]
[letter] :=a..z | A..Z
[d] := 0..9
```

Regarding the Completeness [1] dimension, which evaluates the extent of data that is of sufficient breadth, depth, and scope for the task at hand [22], the following BNF rules can be used to specify a DQSLA:

```
Compl_rules = {[Value_Completeness], [Tuple_Completeness],
  [Attribute_Completeness], [Relation_Completeness]}
[Value_Completeness] := [θ] [d].[d]
[Tuple_Completeness] := [θ] [d].[d]
[Attribute_Completeness] := [θ] [d].[d]
[Relation_Completeness] := [θ] [d].[d]
[θ] := > | < | ≥ | ≤ | ≠ | =
[d] := 0..9
```

Using the above rules, we provide the following example of a DQSLA using the Completeness dimension:

DQSLA_02 = < 022, Completeness, {< 0.9, < 0.9, < 0.9, < 0.9}, (2014-11-09 11:20, 2014-11-11 11:20), 250, 5 min, Online, $SLA_{penalties}$, true >

Similar to the DQSLA_01 semantics, the DQSLA_02 SLA specifies that the data set (022) must be continuously monitored for completeness evaluation during the timestamp interval (2014-11-09 11:20, 2014-11-11 11:20). The evaluation of the monitored data set, which cannot take more than 5 min to be executed, must be reported via a web interface (R_{method} = Online) every time any measure of the completeness dimension is below 90 %, as specified by the M_{rules} content. Besides the initial evaluation of the data set ($Init_{exec}$ = true), the data set (022) must be reevaluated after each 250 changes on its content. Lastly, there is a penalty for the service provider (SLA$_{penalties}$) if the evaluation process takes more time than 5 min.

In practice, a DQSLA may be easily represented as an XML document, and its structure and content may be validated using an XML Schema whose validation rules are derived from BNF rules and from the DQSLA general structure.

1.3.3 Data Quality-Aware Service Architecture

The adoption of cloud computing for databases and data services introduces a variety of challenges, such as strict SLA requirements and Big Data processing. To leverage elastic cloud resources, scalability has to be a fundamental approach of cloud services [16]. This is particularly true in a cloud computing context, in which a virtual infrastructure may be dynamically allocated by considering both the actual conditions of the computing environment and the customers' QoS requirements that have to be met by the service.

In the DQaS context, the service should have the means to process the DQSLA inputs and estimate the right amount of resources that should be allocated to be able to honor the service commitments specified at the DQSLA. In practice, the following inputs may vary over time: (i) execution time of an data quality algorithm execution (e_1), (ii) the time restriction associated to e_1 ($T_{res}(e)$), (iii) the size of the data sets that need to be monitored, and (iv) the complexity of the data quality algorithms that need to be execute by the DQaS. Hence, the amount of resources needed to honor their customers' DQSLAs may also vary notably over time. In order to tackle the infrastructure provisioning problem, we propose the adoption of provisioning algorithms on the DQaS architecture. In practice, these algorithms may use heuristics or machine learning techniques to adjust the infrastructure over subsequent executions or estimate the infrastructure adjustments based on a training data repository that is constantly updated as the DQaS is executed over time.

Another crucial aspect of a cloud service is its ability to maximize the efficiency on the usage of the available resources while minimizing the provider bill [18]. In other words, a cloud service provider has to meet the following two main requirements: (i) guarantee that the customers' SLA requirements are met and (ii) optimize the resource utilization in meeting the customers' requirements.

The DQaS architecture is shown in Fig. 1.2. Initially, the client of the data quality monitoring outsourcing contract signs a contract with the DQaS provider. In practice, the contract is composed by a set of DQSLA inputs. In step 1, a client, who initially owns one or more cloud databases, submits the inputs of a DQSLA using the service interface. In step 2, the received DQSLA is then validated by the *SLA Validator* module using an XML Schema (step 3). In step 4, if the DQSLA structure and content are properly validated by the selected XML Schema, the DQSLA is stored (step 6) as a set of inputs for a data quality algorithm [16].

Then (step 7), if the DQSLA parameter Init$_{exec}$ is set to true, the *Provisioning Planner* module gathers data from the data quality algorithm (derived from the DQSLA) inputs, from the cloud database metadata, and (optionally) from a training

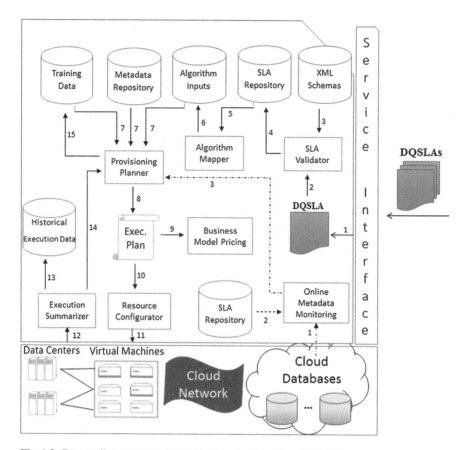

Fig. 1.2 Data quality-aware service architecture (Adapted from Ref. [16])

data repository in order to estimate the ideal infrastructure to be used for the execution of a data quality algorithm. After that (step 8), an execution plan (which consists of a grid configuration) is created. The execution plan specification is used by both the *Business Model* module (to further aggregate customers' bills) and the *Resource Configurator* module (to allocate a grid of virtual machines and execute the data quality algorithm in a distributed manner) [16].

After the execution of the data quality algorithm is completed (step 11 and step 12), the *Execution Summarizer* module aggregates the execution data in order to populate a historical execution data repository (step 13) and also sends the summarized execution data to the *Provisioning Planner* module (step 14) to update the training data used by the service (step 15). Another possible flow (displayed by dashed lines in the DQaS architecture) of tasks performed by the DQaS is triggered when the amount of changes in a data set being currently monitored exceeds the $|\triangle D|_{thr}$ parameter value specified by the client at the DQSLA [16]. When this scenario occurs (which is detected by the *Online Metadata Monitoring* module), the *Provisioning Planner* creates a new execution plan to evaluate the modified data set.

1.4   Future Research Directions

Regarding the presented DQaS architecture (Fig. 1.2), almost every presented module encompasses research and implementation challenges. In this section, each one of these modules and its related research opportunities are separately discussed.

SLA Repository   In this repository, it is necessary to adopt an efficient approach to store and retrieve the customers' DQSLAs according to input parameters. Besides the DQSLA general structure (described in the section "Data Quality SLA Formalization"), other examples of input parameters for querying and retrieving DQSLAs include client id, timestamp of the last triggered execution, and client bill regarding a specific DQSLA. To allow such queries, additional metadata must be associated to the stored DQSLAs. In this context, we identify as development challenges the following activities: (i) the development of a XML-based repository to store and retrieve DQSLAs and (ii) the identification and representation of additional metadata that need to be associated to the DQSLA general structure in order to facilitate the process of querying and retrieval of DQSLAs.

Online Metadata Monitoring   For the implementation of this module, it is necessary to investigate and develop efficient algorithms to calculate the amount of changes that affect cloud databases as the database management system (DBMS) processes incoming transactions. Moreover, this module implementation must take into account many challenges, such as: (i) the heterogeneity of the DBMSs, (ii) the amount of processed data, and (iii) the different types of database logical schemas.

XML Schemas and DQSLA Validator   In the section "Examples of Data Quality SLAs," we provided an example of the M_{rules} inputs by using specific BNF rules for defining parameters regarding both the Duplication and Completeness data quality dimensions. In order to specify further M_{rules} for other data quality dimensions, it is necessary to perform a survey of the state of the art regarding the catalog of data quality dimensions and create specific BNF rules regarding the parameters related to these dimensions in a DQSLA. In this context, it is also necessary to specify and develop XML Schemas that will be used by the DQaS for validating both the structure and content of customers' input DQSLAs. Finally, it is required to develop a DQSLA Validator that is able to select dynamically an XML Schema from the repository, according to the DQ_{dim} value of an input customer DQSLA, and validate both the content and the structure of the input DQSLA. Using this approach, the XML Schema repository may be updated and extended over time without affecting the DQSLA Validator module.

Historical Execution Data   As the DQaS executes over time, it is important to maintain a historical record of the executions of data quality algorithms. First, the output of such algorithms consists of aggregated measures concerning the quality of the monitored databases and thus may be interesting for real-time and historical analysis. Second, the results of the data quality algorithm executions, such as execution time and costs, should be easily queried for customers' analytical analysis. In this sense, data warehousing techniques may be used in order to model a multidimensional database that can be used to store subject-oriented, temporal, and nonvolatile data generated by the execution of data quality algorithms over time.

Penalty Model In the DQaS context, it is necessary to adopt a penalty model to calculate the amount of penalty that is inflicted to the DQaS for the cases in which the service does not meet the time restriction, specified in the DQSLA, for the execution of a data quality algorithm. For doing so, regarding a data quality algorithm execution e_1 triggered by a DQSLA that defines a time restriction equal to $T_{res}(e_1)$, the DQaS penalty model must take into account (i) the difference between the execution time of e_1 and $T_{res}(e)$; (ii) the predefined DQaS penalty parameters, such as the fixed penalty and the penalty rate; and (iii) the penalty function that is used to apply the penalty rate. For example, let α be the fixed penalty and β the penalty rate. Using a linear penalty function, one can calculate the total penalty cost as follows: $\alpha + \beta$ (ExecTime$(e) - T_{res}(e)$). The related literature discusses many different approaches [23] for a penalty model. Thereby, these approaches need to be investigated and evaluated in order to conclude which penalty model is more suitable for the DQaS context. Alternatively, one can also propose and evaluate novel penalty models that are designed specifically for a DQaS.

Cost Model As stated earlier in this book chapter, one of the main goals of a service provider in the cloud computing context is to minimize the service costs (and, in turn, customers' bill) by meeting customers' requirements (predefined QoS) using the minimum amount of hardware infrastructure. In this context, in order to measure hardware-level infrastructure costs, it is necessary to design mathematical cost models that need to take into account the following requirements: (i) evaluation of the infrastructure costs for the execution of data quality algorithms using a cluster of virtual machines, (ii) evaluation of infrastructure costs related to the cloud maintenance, (iii) penalty costs, and (iv) aggregation of the different types of costs in order to calculate the DQaS total cost.

Charging and Profit Models The key assumption for the deployment of a DQaS is its capability of providing a cloud service whose total cost (that is aggregated from hardware, maintenance, and penalty costs) is inferior to the aggregated customer's bill. For doing so, it is necessary to develop and evaluate charging estimation algorithms, in the *Business Model Pricing* module of the DQaS architecture, which are able to provide a good estimation for a customer's bill, given (i) the customer DQSLA inputs, (ii) the estimated maintenance and infrastructure costs required to meet the expected QoS, (iii) the expected penalties over time, and (iv) the desired profit margin for the DQaS. Regarding the business model perspective of the DQaS, the charging and the profit models are particularly crucial for the service, since low-quality charging estimations may lead to losses (i.e., a charged bill that is inferior than the service total cost) or lack of competitiveness (i.e., a charged bill that is much superior than the service total cost) for the service.

Provisioning Algorithms In order to tackle the infrastructure estimation problem, the DQaS architecture uses a module (*Provisioning Planner*) that enables

DQSLA-driven dynamic configuration, management, and optimization of cloud resources. Using these functionalities, the DQaS may respond effectively to the QoS requirements of the service customers. As the computational demands for the executions of data quality algorithms fluctuate, the allocated hardware infrastructure and service pricing must be adjusted accordingly. The success of this process derives from the provider's ability to profile the service's resource demands, provision those demands with sufficient resources, and react as resource supply and demand change [24]. For doing so, it is necessary to develop provisioning algorithms for the *Provisioning Planner* module that are able to (i) estimate an initial virtual machine infrastructure for the first execution of a data quality algorithm triggered by a customer DQSLA and (ii) adjust the virtual machine infrastructure after subsequent executions of data quality algorithms triggered by the same customer DQSLA. In summary, it is expected that, for each data quality algorithm execution *e*, the *Provisioning Planner* should allocate a virtual machine cluster that is able to approximate ExecTime(e) $\approx$ T$_{res}$(e). To this end, the provisioning algorithms may use or combine consolidated artificial intelligence techniques, such as heuristics and machine learning. The former technique may be used to adjust the allocated infrastructure over subsequent executions, whereas the latter may be used to provide an initial virtual infrastructure estimation for the first execution triggered by a customer DQSLA. The machine learning approach needs to use a training data repository, which is initially filled by data provided by executions of data quality algorithms in the cloud and is constantly updated as the DQaS executes over time. Thereby, the more the DQaS executes over time, the better are the virtual infrastructure estimations provided by the *Provisioning Planner*. Using the described training data repository, the machine learning provisioning algorithms may perform the following tasks: (i) use predefined measures to calculate the similarity between the inputs of a given execution triggered by a DQSLA and the records of the training data repository, (ii) order the records according the calculated similarities, and (iii) select the virtual machine infrastructure based on the computed similarities and a classification model.

Efficient Techniques for Data Quality Algorithms Another approach that can be used by the DQaS in order to minimize infrastructure costs (and, in turn, decrease customers' bill as well as improve service competitiveness), specially for handling Big Data, is the adoption of efficient techniques for the data quality algorithms. This general guideline encompasses many different approaches, such as efficient distributed execution, the usage of indexing techniques, and the adoption of incremental data quality algorithms. For example, regarding the deduplication task [8], the following guidelines may be used:

- *Efficient distributed execution*: the DQaS must use consolidated approaches [12, 13, 25, 26] for the distributed execution of data quality algorithms, such as the MapReduce [27] paradigm. Besides, the service must also adopt strategies for tackling specific problems related to distributed executions, such as load unbalancing [8]. This problem is experienced when unbalanced tasks are distributed among the executing nodes, since the overall execution time of a data quality algorithm will be bounded by the execution time of the last node that completes its task. In practice, this problem may happen due to data skewness or ineffective task allocations.

- *Indexing techniques*: the performance bottleneck in a deduplication system is usually the expensive detailed comparison of field (attribute) values between records [28], making the naive approach of comparing all pairs of records not feasible when the databases are large [8]. Moreover, the vast majority of comparisons will be between records that are not matches (i.e., true duplicated). Thereby, a good approach to reduce the amount of required comparisons is the adoption of an indexing technique called *blocking* [14]. This technique consists in dividing the database records into nonoverlapping blocks, which are generated based on the values of a *blocking key* of each record, such that only the records that belong to the same block need to be compared among each other. Thus, it is important to adopt such indexing techniques to optimize the execution of the data quality algorithms in the DQaS context. The DQaS should also aid the customers to estimate parameters related to the indexing techniques, such as the blocking key of a data set and appropriate thresholds for deduplication algorithms.
- *Incremental deduplication*: in the Big Data era, the velocity of data updates is often high, quickly making previous linkage results obsolete [10]. This challenge calls for an incremental strategy for reprocessing the previous deduplication results when data updates arrive. In this sense, it is necessary to explore the related literature in order to incorporate such incremental techniques as part of the DQaS optimization approaches.

Evaluation of the DQaS in the Cloud   Before being tested in a real-world scenario, the overall functioning of the presented DQaS architecture must be validated in a real cloud computing environment. These initial tests are crucial to validate the effectiveness of some of the key modules of the presented architecture, such as the *Provisioning Planner* and the *Business Model Pricing*. The service evaluation may be initially carried out through simulation. Mathematical models and simulation environments may be developed to simulate the execution of the data quality algorithms in a cloud computing infrastructure. Such models or environments must take into account a number of factors, such as: (i) the computational complexity of the data quality algorithms, (ii) the influence of the data set size over the execution time of the algorithms, (iii) the influence of the virtual infrastructure processing capacity over the processed tasks, and (iv) the influence of external factors, such as network delays, data reading, and the load balancing problem. The results of the simulation exercise are important to provide initial insights about the effectiveness of different provisioning algorithms and opportunities for optimizations on the service, such as choosing tunable parameters for the *Business Model Pricing* module. After initial evaluations through simulation, further efforts are required to validate the DQaS architecture modules in a real cloud computing environment. Ideally, these experiments must evaluate the total cost to provide this service using real workloads, i.e., composed by real data sets and data updates, and the ability of the service to perform effective adjustments of the allocated virtual infrastructure as the QoS demands (specified in the customers' DQSLA inputs) fluctuate over time. These experiments should also be performed using Big Data data sets in order to evaluate the DQaS capabilities for handling the scale of existing contemporary databases. As a result, these experiments will provide important insights regarding appropriate provisioning algorithms and mathematical models (for service cost, pricing, and profit) for the DQaS. These results are also crucial to determine the applicability of the presented architecture as a real-world business model.

1.5   Conclusions

In this book chapter, we explain the formalized notion of a data quality SLA that can be used to describe expectations and obligations of the both parts (*customers* and *service provider*) involved in a data quality monitoring outsourcing contract. We then discussed a number of challenges that encompass the development of a data quality-aware service, mainly those related to the definition of QoS in the data quality context and the dynamic allocation of virtual infrastructure to tackle: (i) customer's QoS demands and fluctuations and (ii) Big Data processing. In order to tackle the described problems, we have discussed a high-level DQaS architecture. Its main goal is to meet the customer's requirements, specified as DQSLA inputs, by minimizing infrastructure costs and penalties over time.

Due to the cloud nature of the DQaS architecture, we have shown that both the deployment of such service and the development of its main modules encompass many research and technical challenges. Database practitioners, professionals, and researchers may further tackle these challenges by using existing techniques and technologies or by proposing and evaluating new approaches. Finally, the proposed evaluation of the DQaS, using a real cloud computing scenario, is expected to provide encouraging results. If so, the service may be further explored commercially as a real-world business model.

References

1. Loshin D (2010) The practitioner's guide to data quality improvement. Elsevier, Burlington
2. Sadiq S (ed) (2013) Handbook of data quality. Springer, New York
3. Buhl HU, Röglinger M, Moser DK, Heidemann J (2013) Big data: a fashionable topic with(out) sustainable relevance for research and practice? Bus Inf Syst Eng 5(2):65–69
4. Kaisler S, Armour F, Espinosa JA, Money W (2013) Big data: issues and challenges moving forward. In: Proceedings of the 46th Hawaii international conference on system sciences (HICSS), pp 995–1004
5. Katal A, Wazid M, Goudar RH (2013) Big data: issues, challenges, tools and good practices. In: Proceedings of the 6th international conference on contemporary computing, pp 404–409
6. Badidi E (2013) A cloud service broker for SLA-based SaaS provisioning. In: Proceedings of the international conference on information society, pp 61–66

7. Schnjakin M, Alnemr R, Meinel C (2010) Contract-based cloud architecture. In: Proceedings of the second international workshop on cloud data management, pp 33–40
8. Christen P (2012) A survey of indexing techniques for scalable record linkage and deduplication. IEEE Trans Knowl Data Eng 24(9):1537–1555
9. Bizer C, Boncz P, Brodie ML, Erling O (2012) The meaningful use of big data: four perspectives – four challenges. ACM SIGMOD Record 40(4):56–60
10. Gruenheid A, Dong XL, Srivastava D (2014) Incremental record linkage. Proc VLDB Endowment 7(9):697–708
11. Ioannou E, Rassadko N, Velegrakis Y (2013) On generating benchmark data for entity matching. J Data Semantics 2(1):37–56
12. Hsueh SC, Lin MY, Chiu YC (2014) A load-balanced mapreduce algorithm for blocking-based entity-resolution with multiple keys. In: Proceedings of the 12th Australasian symposium on parallel and distributed computing, pp 3–9
13. Mestre DG, Pires CE, Nascimento DC (2015) Adaptive sorted neighborhood blocking for entity matching with mapReduce. In: Proceedings of the 30th ACM/SIGAPP symposium on applied computing, pp 981–987
14. Baxter R, Christen P, Churches T (2003) A comparison of fast blocking methods for record linkage. ACM SIGKDD 3:25–27
15. Dillon T, Wu C, Chang E (2010) Cloud computing: issues and challenges. In: Proceedings of the 24th IEEE international conference on advanced information networking and applications, pp 27–33
16. Nascimento DC, Pires CE, Mestre D (2015) A data quality-aware cloud service based on metaheuristic and machine learning provisioning algorithms. In: Proceedings of the 30th ACM/SIGAPP symposium on applied computing, pp 1696–1703
17. Dan A, Davis D, Kearney R, Keller A, King R, Kuebler D, Youssef A (2004) Web services on demand: WSLA-driven automated management. IBM Syst J 43(1):136–158
18. Ferretti S, Ghini V, Panzieri F, Pellegrini M, Turrini E (2010) Qos–aware clouds. In: Proceedings of the IEEE 3rd international conference on cloud computing, pp 321–328
19. Skene J, Lamanna DD, Emmerich W (2004) Precise service level agreements. In: Proceedings of the 26th international conference on software engineering, pp 179–188
20. Batini C, Cappiello C, Francalanci C, Maurino A (2009) Methodologies for data quality assessment and improvement. ACM Comput Surv 41(3):1–52. doi:10.1145/1541880.1541883, ISSN: 0360–0300
21. Sidi F, Shariat PH, Affendey LS, Jabar MA, Ibrahim H, Mustapha A (2012) Data quality: a survey of data quality dimensions. In: Proceedings of the international conference on information retrieval and knowledge management, pp 300–304
22. Wang RY, Strong DM (1996) Beyond accuracy: what data quality means to data consumers. J Manag Inf Syst 12(4):5–33
23. Rana OF, Warnier M, Quillinan TB, Brazier F, Cojocarasu D (2008) Managing violations in service level agreements. In: Grid middleware and services. Springer, pp 349–358. http://link.springer.com/chapter/10.1007/978-0-387-78446-5_23
24. Reynolds MB, Hopkinson KM, Oxley ME, Mullins BE (2011) Provisioning norm: an asymmetric quality measure for SaaS resource allocation. In: Proceedings of the IEEE international conference on services computing, pp 112–119
25. Kolb L, Thor A, Rahm E (2013) Load balancing for mapreduce-based entity resolution. In: Proceedings of the IEEE 28th international conference on data engineering, pp 618–629
26. Mestre DG, Pires CE (2013) Improving load balancing for mapreduce-based entity matching. In: IEEE symposium on computers and communications, pp 618–624
27. Dean J, Ghemawat S (2008) MapReduce: simplified data processing on large clusters. Commun ACM 51(1):107–113
28. Christen P, Goiser K (2007) Quality and complexity measures for data linkage and deduplication. In: Quality measures in data mining. Springer, Berlin/Heidelberg

Chapter 2
Role and Importance of Semantic Search in Big Data Governance

Kurt Englmeier

Abstract Big Data promise to funnel masses of data into our information ecosystems where they let flourish a yet unseen variety of information, providing us with insights yet undreamed of. However, only if we are able to organize and arrange this deluge of variety according into something meaningful to us, we can expect new insights and thus benefit from Big Data. This chapter demonstrates that text analysis is essential for Big Data governance. However, it must reach beyond keyword analysis. We need a design of semantic search for Big Data. This design has to include the individual nature of discovery and a strong focus on the information consumer. In short, it has to address self-directed information discovery. There are too many information discovery requests that cannot be addressed by mainstream Big Data technologies. Many requests often address less spectacular questions on a global scale but essentially important ones for individual information consumers. We present an open discovery language (ODL) that can completely be controlled by information consumers. ODL is a Big Data technology that embraces the agile design of discovery from the information consumer's perspective. We want users to experiment with discovery and to adapt it to their individual needs.

2.1 Introduction

Big Data promise to funnel masses of data into our information ecosystems where they let flourish a yet unseen variety of information, providing us with insights yet undreamed of. However, we have to organize and arrange this deluge of variety into something meaningful for those that expect new insights when consuming this new variety of information.

Following the actual discussion, we have to address some essential issues if we want to reap the benefits from Big Data. They result from better customer relationships, improved market insights, and better responsiveness in health care, just to name a few. To get there, we have to revamp our information strategy; we

K. Englmeier (✉)
Faculty of Computer Science, Schmalkalden University of Applied Science, Blechhammer 4-9, 98574 Schmalkalden, Germany
e-mail: KurtEnglmeier@computer.org

© Springer International Publishing Switzerland 2015
M. Trovati et al. (eds.), *Big-Data Analytics and Cloud Computing*,
DOI 10.1007/978-3-319-25313-8_2

require a data-driven mind-set; quite likely we need new personnel, a chief data officer, and, yes, Hadoop, MapReduce, and many more analytics tools. Only a powerful mix of all these ingredients prevents us from being drowned by the Big Data avalanche.

However, powerful analysis machines help us to write only half of our Big Data success stories, at most. The real challenge in Big Data is the discovery of the right data in masses of unstructured data. More than our analysis capabilities Big Data challenges our data discovery capabilities. With the vast majority of Big Data being text data, information discovery in texts gained a new momentum in Big Data research.

This chapter demonstrates that text analysis is essential for Big Data governance. However, it must reach beyond keyword analysis. We need a design of semantic search for Big Data. This design has to include the individual nature of discovery and a strong focus on the information consumer. Hence, information demand depends on the interests of the consumers. The core element of the design is a common vocabulary that abstractly outlines the information ecosystem the respective community of information consumers is dealing with. The common vocabulary enables self-service discovery by supporting semantic search that is completely controlled by the consumer. They compose blueprints of their information requests that are finally machine processable to perform automatic discovery. This article proposes a machine-processable language the information consumers can apply to manage their self-service discovery.

2.2 Big Data: Promises and Challenges

The idea behind Big Data analytics is to drill in mountains of data with powerful and sophisticated tools in hopes of unearthing *new insights*. However, is capitalizing on Big Data just a matter of better machines? Can we effectively tackle the Big Data issue with better database performance, more statistical tools, enhanced data mining (DM) methods, and the like? Of course, Big Data means more data, including input from sensors. The cell phone produces position data. Location-aware browsers and apps log when and where you state a query or use certain features. Smart watches monitor your heart rate and your pulse, among other things. These examples demonstrate that there are in fact new interesting data to digest for analytic tools. It is probably the advent of these sensor data that makes us believe that capitalizing on Big Data is tantamount to enhanced and enriched number of crunching capabilities.

Besides some new types of data, there is not so much new in Big Data analysis. DM and information retrieval (IR) models and methods are applied to a new variety of data. DM and IR can produce, in fact, interesting and valuable information. It is certainly reasonable to apply well-established and acknowledged methods from artificial intelligence (AI). The recurring theme in Big Data is *correlation*. The correlation of the phone's position data with the use of special keywords revealing that we might suffer a flu or the like can indicate where and how fast a disease spreads. The use of an inhalator correlating with position data may reveal areas problematic for people suffering from asthma.

Today, one main focus in Big Data research is on automatically detecting correlations between data of different nature, pulse rate with position data, or restaurant descriptions with position data, for instance. Some new insights are quite voyeuristic like in the example of the Uber taxi service that claimed to have detected service usage patterns indicating presumable one-night stands of its clients [14]. To presume that the detected correlation points to just this one cause is both embarrassing and unscrupulous. However, it sheds light on a very big problem Big Data analysis has: false positives. Quite often, Big Data analysis does not investigate any further the broad variety of causes as long as correlations indicate new and spectacular insights. A further example is the prognosis of the spread of a flu epidemic based on tweets combined with their authors' position data. Here, Big Data analysis pretended to detect velocity and directions of the spread by identifying flu-related keywords in tweets that were linked with the location where the users posted these tweets. Later, the results of this analysis were shown to be somewhat far from reality [7]. In the first place, the two examples demonstrate the *threats of Big Data* if we trust too much in the authority of the analytic tools and neglect a thorough analysis of the causes that produced the phenomena appearing in the correlations. This may lead to severely false conclusions when we take the produced phenomena as facts too fast. The two examples also demonstrate that a close involvement of humans is indispensable if we want a sustainable success in Big Data that reaches beyond making some money based on hype.

Big Data stories tell about new insights that result from the application of analytic algorithms on data available in databases or made available in databases by tools like MapReduce. Data analysis, in particular high-volume analysis, depends on suitable abstract data descriptions, this means, on machine-processable data schemata. The correct identification of the nature of data is essential for successful analytics. A database schema clearly indicates this nature, and it clearly marks sales figures, zip codes, locations, and the like as such. For decades, we shovel these data from all kind of applications (e-commerce, CRM, etc.) into our databases. These applications make sure we get all these data in SQL compatible format. Today, we have, in addition, Big Data tools like Hadoop that help us to extract some SQL incompatible data, mainly from sensor data, and convert them into digestible bites for our databases. This in turn means even the most sophisticated analytic tool does not help much if we cannot feed it with suitably formatted data. There are no insights without suitably structured data.

2.3 Participatory Design for Big Data

Big Data analysis resembles much that we already know from AI since decades, in particular from DM and IR. Correlations in data may stand for relationships between facts pointing to a phenomenon. If I'm using keywords related to a certain disease, I'm suffering from this disease and require appropriate treatment. We see immediately that the observed phenomenon is first of all an assumption about both the person and the illness. And these assumptions can be wrong. I may look up this information for somebody else, for instance. The combination of keywords may also stand for a completely different phenomenon.

The foundation of these assumptions is the analysis models. However, the verification of these models is barely part of model design. Usually, these models emerge from statistical methods, from factor or time series analysis, or are based on hidden Markov models, to name the most prominent areas. As long as they detect a critical mass of correlations, the problem of false positives does not gain sufficient momentum.

"We need to avoid the temptation of following a data-driven approach instead of a problem-driven one" [5]. *Information consumers* can usually sketch their information demand that summarizes the data they need to solve their information problem. They have a *deep understanding of the foundations of their domain.* Thus, we need a stronger involvement of humans in the value chain of Big Data analysis. Sustainable success in Big Data, however, requires more than just controlling the results produced by Big Data analytics. The integration of more competence, in particular domain competence, means a more active role of the human actor in all stages of the value chain. This kind of user engagement goes beyond user requirement analysis, participatory design, and acceptance testing during the development of Big Data analytic systems. It means a more active role of the information consumers enabled by self-service features. This kind of self-service IT may point to user-friendly versions of analytic tools, enabling information consumers to conduct their own analytics. This smooth integration of domain and tool knowledge completes the picture of *self-service discovery* that meanwhile is also demanded by the industry [12]. There is no doubt that without proper technological and methodological support, the benefits of Big Data are out of reach. Design for Big Data, however, requires an engagement of information consumers. When their needs drive design, Big Data will provide the insights they require. It is too often the case that technology as required and described by the users is not quite well understood by designers. There are several ways of human-centered design to overcome this lack of mutual understanding. There is user-centered design where users are employed to test and verify the usability of the system. *Participatory design* [11] makes available this form of user engagement; it understands users as part of the design team.

An organization's information ecosystem usually hosts databases, content management systems, analytic tools, master data management systems, and the like that produce the flavor of information specific to this organization. Furthermore, there are communication tools, CRM and BPM systems, that feed the ecosystem with data, too. When we talk about information systems, we mainly mean systems and the data they manage. The users see these systems as different data channels that provide different blends of information. With the help of IT, they can retrieve their information through these channels, or they request reports containing the required information. For the users, these channels have precise qualities in terms of the information they expect to get. There is a database containing "household data, including household income, age structure, education levels of its members, etc.," for example, or one that hosts "reports from the radiology." Users know these sources by an abstract description of the data they contain; they annotate keywords to these sources giving them a semantic badge. Of course, every user may assign a slightly different badge to the same data source, but in general the assignments are clear to all users. In the same way, users address the analytic tools and features an organization provides. In any organization there exists a quite informal, abstract semantic representation of shared information.

If we approach information governance from the information consumer perspective, it is probably a good idea to take these semantics as building blocks for an information governance framework.

Likewise, by assigning a set of keywords to every data channel, we can assign similar semantic badges to any concept reflected in the data of this channel. If we link all semantic badges over all data channels, we get a complete semantic representation of the information ecosystem the information consumers are dealing with. The network of semantic badges communicates what information the users can expect.

The management of such a semantic representation is quite a human endeavor. They author suitable keywords representing the concept that constitutes particular information. The set of keywords thus acts as a blueprint for particular information. At first, the badge is an abstract representation of information, written in natural language. They are not machine processable at this stage. A contract may have a badge labeled with the concepts "vendor," "buyer," "product," "date," and "price." The badge assigned to a diagnosis of a patient may list the key concepts "name," "age," "observations," and "date." Each of these concept representations can be composed of further concepts. The concept "product" may stand for a "property" described by "location," "size," and further attributes like the building erected on it. The observations stated in the medical report can be further detailed along the organs they address and typical phenomena, like "hemorrhage," "lump," etc. The concept "organ" again can be further detailed into its constituent parts and so on. In the end, we get a semantic representation of concepts that resembles a thesaurus. However, it is not a general thesaurus, rather an individual one, adapted to the specifics of the respective information ecosystem. Furthermore, it isn't either a strict thesaurus where all its concepts are tightly integrated. It's rather a collection of more or less loosely coupled fractions of a thesaurus, with its fractions dynamically changing both, in their compositions and relationships among each other. It is thus more suitable to consider semantic badges as ingredients of a common vocabulary. This vocabulary, in turn, is the asset of the information consumers. They manage it in cooperative authorship.

People working with information have a data-driven mindset per se [10, 15], that is, they resort to mental models [9] that abstractly reflect the facts they expect to encounter in their information environment [1]. This mindset enables them to sketch blueprints of the things they are looking for. Information consumers can express these blueprints in a way that later on can be processed by machines.

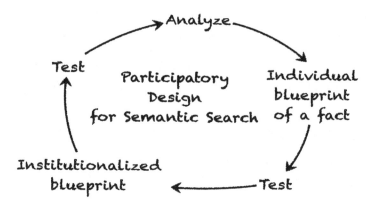

Fig. 2.1 Design cycle for the cooperative development of blueprints that govern the discovery process

These expressions are far from being programming instructions but reflect the users' "natural" engineering knowledge [16]. The machine then takes the blueprints and identifies these facts in data, even though the blueprint abstracts away many details of the facts.

Experimenting with data is an essential attitude that stimulates *data discovery experiences*. People initiate and control discovery by a certain belief – predisposition or bias reinforced by years of expertise – and gradually refine this belief. They gather data, try their hypotheses in a sandbox first, and check the results against their blueprints, and then, after sufficient iterations, they operationalize their findings in their individual world and then discuss them with their colleagues. After having thoroughly tested their hypotheses, information consumers institutionalize them to their corporate world, that is, cultivate them in their information ecosystem. After reflecting the corporate blueprint in their individual world, they may get a further idea for further discoveries, and the participatory cycle starts anew (Fig. 2.1).

The language knowledge and their mental models constitute shallow knowledge necessary and sufficient to engineer statements that are processable by the discovery services [13]. The blueprints thus serve two purposes: they reflect semantic qualities of the facts that need to be discovered, and simultaneously, they are the building blocks of the metalanguage that, when correctly syndicated, support data integration and sharing. While syndicating metadata along their domain competence, users foster implicitly active compliance with organizational data governance policies.

2.4   Self-Service Discovery

There are many designs for discovery languages that are rooted in semantic search. Usually, they apply Semantic Web languages like RDF, OWL, and the like [3]. Usually, these languages turn out to be too complicated in order to be handled by information consumers. What is required is an *open discovery language (ODL)* that is designed along the paradigm of simplicity in IT [8]. The users shall concentrate on discovery and sharing, stick to their data-driven mind-set, and express their requests as far as possible in their own language.

A good starting point therefore is the language of information consumers. To avoid any irritations or ambiguities, people try to be quite precise in their descriptions. Even though these descriptions are composed of natural language terms and statements, humans are quite good in *safeguarding literal meaning* in their descriptions [6]. We avail ourselves of literal meaning because we can interpret statements correctly in the absence of any further explicit and implicit context. This aspect is also important when it later comes to machine interpretation of these statements. They operate on data enriched by annotations that make implicit meaning explicit. They also set up the semantic skeleton of the information ecosystem where search and discovery happens. This skeleton represented as a network of metadata emerges from the domain knowledge of the information consumers. For us, the data-driven mind-set becomes evident when users can turn their domain and shallow engineering knowledge into machine instructions suitable for the precise detection and extraction of the facts they expect to discover. The blueprints thus serve two purposes: they reflect semantic qualities of the facts the discovery engine shall locate and extract, and simultaneously, they are the building blocks of the metalanguage that, when correctly syndicated, support data integration and sharing. While syndicating metadata along their domain competence, users foster implicitly active compliance with organizational data governance policies.

Information discovery starts with *information extraction* (IE) [2] that distills text or even scattered documents to a germ of the original raw material. IT experts engineer information extraction systems that operate on templates for the facts to be extracted. Labeled slots constitute these templates whereby the labels represent annotated terms.

Self-service information discovery starts with user-driven IE. The users first have to engineer their extraction templates that can also be considered as entity recognizers. This means a certain amount of engineering is indispensable in IE. The key question is whether information consumers have the necessary engineering knowledge to handle discovery services on their own. This shallow knowledge is assumed to be acquired easily and is thoroughly specific to the task at hand [4]. The assumption in self-service discovery is that users with their domain competence and shallow engineering knowledge are in the position to manage a certain level of data discovery and information sharing on their own.

The users' blueprints may be simple and concise, but they are comprehensive enough to cover their request. This, in turn, fosters the control of the discovery process. A template with, say 8–12 slots, can comprehensively cover real-world requests in small-scale domains. This low level of complexity makes it easy for the information consumer to manually control discovery. Whenever they encounter unfilled slots or mistakenly filled slots, they may check the corresponding document for obvious errors. On the other hand, they may adapt their blueprint if the representation of a fact appears to be sound in text, but the slots do not consistently correspond to the qualities of the fact.

IE knows a lot of methods to match text excerpts (i.e., candidates for information entities) with slot keys (i.e., labels) and to fill the slots. In our approach, we focus on pattern recognition. Each template slot is thus linked to a descriptive pattern that is rendered as Regular Expression in combination with key terms. Regular Expressions are a powerful instrument to precisely detect all kinds of patterns in data.

Their application is in particular useful for the detection of facts in unstructured information.

Furthermore, they offer the opportunity to sketch a variety of patterns for a particular fact that may appear in many variant forms. A date, for instance, can be expressed in different forms even within the same language. Regular Expressions have a particular advantage, too, if data represent a named entity, e.g., as combination of numerical data and words. A birthdate may be rendered by keywords ("born on") or symbols (an asterisk) forming a specific pattern with an adjacent date. The position within text may also suffice to qualify a date as birthday: "Franz Kafka (3 July 1883–3 June 1924)." In each text, we can identify numerous facts of different complexities. Many of them can be identified in patterns that comprise facts located in close proximity. Information on a person may span over more basic facts like name, taxpayer number, birthdate, address, etc. Some fact may comprise constituent facts that are widely scattered over a broader area of the data, even over more text pages or documents, for instance.

The overarching objective of an *ODL* must be high level of *usability*, in particular in terms of learnability, memorability, and ease of use. Information consumers have a *shared mental model* about appearance and nature of descriptive (word) patterns: entities corresponding to an information blueprint can be dispersed over a number of text blocks (facts), for instance, [vendor] sells [object] to [buyer]. Within each block, terms may appear in arbitrary order, like the characteristics of a person or organization. Some terms are optional, and there are keywords like "sell" or "buy" that essentially characterize the fact "transaction purchase." The question then was how the appearance of terms and term blocks can be expressed by operators.

Regular Expressions are a powerful but absolutely not a user-friendly instrument. They require special skills and are not easy to handle, in particular, when they are addressing complex, i.e., real-world, patterns. Besides, Regular Expressions representing high-level facts are extremely complex and barely manageable, even for professionals. Their application also has limitations, when relevant elements (qualities) of facts are too dispersed over the data set that means when too much "noise" appears between facts and their qualities. We therefore propose a language for discovery and sharing that adopts Regular Expressions but shields users from their complexity. IT provides users with a stock of labeled Regular Expressions addressing entities like "word," "taxpayer number," "phrase," "decimal," etc. (see Fig. 2.2). Instead of defining Regular Expressions, the users compose their patterns by resorting to these basic patterns or to the patterns they have already defined by themselves. The ODL syntax serves to describe how facts, as the constituent parts of the fact requested by the user, appear as sequences of word patterns in data. The

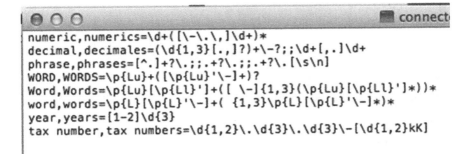

```
numeric,numerics=\d+([\-\.\,]\d+)*
decimal,decimales=(\d{1,3}[.,]?)+\-?;;\d+[,.]\d+
phrase,phrases=[^.]+?\.;;.+?\.;;.+?\.[\s\n]
WORD,WORDS=\p{Lu}+([\p{Lu}'\-]+)?
Word,Words=\p{Lu}[\p{Ll}']+([ \-]{1,3}(\p{Lu}[\p{Ll}']*))*
word,words=\p{L}[\p{L}'\-]+( {1,3}\p{L}[\p{L}'\-]*)*
year,years=[1-2]\d{3}
tax number,tax numbers=\d{1,2}\.\d{3}\.\d{3}\-[\d{1,2}kK]
```

Fig. 2.2 Examples of basic patterns (entity recognizer). Usually, IT experts provide these patterns to information consumers who construct their patterns from here

corresponding descriptive pattern gradually aggregates into the complex pattern for the requested fact.

The users achieve the definition of complex descriptive patterns by iteratively aggregating their pattern definitions starting from the basic patterns (see Fig. 2.2) provided by their IT colleagues. This means any pattern (if not a basic pattern) can be decomposed into smaller units from top to bottom, eventually into basic patterns and key terms. The discovery service sees any pattern as more or less big Regular Expression. These patterns at the bottom layer act as "connector patterns" because they link to the respective data representation technique, used by the service, in our case Regular Expressions. They resemble primitive data types. The syntax for the definition of patterns is quite simple. It supports a handful of operators the users employ to define a descriptive pattern as a sequence of constituent elements, i.e., word patterns.

The following generic definition of a descriptive pattern summarizes the syntax of ODL for discovery and integration; Table 2.1 explains in more detail the functionality of the operators:

```
concept, concepts = element1.element2; element3.
                    (element4,element5):name.?element6
```

On the left side of the equation, the user defines the name of the pattern. The right side of the equation lists the sequence of pattern elements. The pattern can be assigned to a singular term and optionally also to its corresponding plural term. When defining their own patterns, people intuitively apply both forms.

The following application of ODL on the (English) Wikipedia collection shall illustrate features and functionalities of the discovery language. Let us extract birth- and death dates of German writers and to list their works with their original title and the corresponding English translation of the title. By checking a sample of Wikipedia pages, we familiarize ourselves with the way how these facts are represented (Fig. 2.3) and defined from the corresponding descriptive patterns using ODL (Fig. 2.4). Applying the blueprints on this collection returns the extracted information in XML format (Fig. 2.5).

Table 2.1 Operators of the open discovery language (ODL)

Operator	Function
.	The dot indicates strong sequence ("followed by"). The fact indicated before the dot must be located before the one indicated after the dot
,	Comma means weak sequence. Some of indicated facts (at least one) shall appear sequentially in the data. However, they may appear in any order (inclusive combination)
;	The semicolon is used to indicate an exclusive combination. Just one of the facts ought to be located
:	Labeling operator: the name after the colon is assigned to facts or a group of facts. Labeling serves the implicit introduction of (local) nested patterns
(. . .)	Parentheses serve to indicate a group of facts. Grouping only makes sense together with the labeling operator
?	The question mark indicates that a fact or group of facts can be optional, that is, the corresponding fact can but need not be located in the data sources
"Keyword"	The discovery service treats keywords indicated between quotation marks as fixed expressions. If it represents a verb, the keyword is reduced to its principal part. Each noun or adjective is expanded to a Regular Expression covering their possible grammatical forms (flections)
	Three dots (. . .) within a constant indicate that there may appear a small number of irrelevant terms between the elements of the fixed expressions

The example illustrates also the rationale for data integration and information sharing behind ODT that supports the collaborative development and institutionalization of a metadata schema. This schema can constitute the semantic skeleton of an information ecosystem on group or organizational level. In this context, self-service discovery and integration support "active compliance," that is, the collaborative agreement on a unified overarching metadata schema.

As already said, self-service discovery addresses, among other things, ad hoc requests for discovery that, like in the example above, are not too complex. The integration in terms of consolidating metadata schema on a broader organizational level is thus not the only issue in integration. Many blueprints for small-scale and ad hoc requests are disposable artifacts for individual purposes. The integration into the technical environment is more important for discovery requests that need to be shared. What kind of input formats is supported by the discovery service, and how are the results presented? For the time being, our service accepts input such as HTML, PDF, or plain text documents. The output is simply rendered in XML, in order to enable a smooth integration into follow-up processes for data analytics, visualization, reporting, and the like. The slots (with annotated terms) capture the facts according to the users' pattern definitions render them by XML elements. Basic patterns are treated like primitive data types; entity elements that correspond to them are not explicitly tagged.

The first experiments with the ODL outlined here have been performed. They addressed about 2000 documents (real estate contracts with related certificates) distributed over 18 data sources and about 180 diagnosis reports from radiology.

Heiner Müller (January 9, 1929 – December 30, 1995) was a German
Described as "the theatre's greatest living poet" since Samuel Beckett,
after Bertolt Brecht. His "enigmatic, fragmentary pieces" are a significar

however, with his drama *Die Umsiedlerin* (*The Resettler Woman*)
from the Writers' Association in the same year. The East Germar
niere of *Der Bau* (*Construction Site*) in 1965 and censoring his *Ma*

Franz Kafka[a] (3 July 1883 – 3 June 1924) was a German-language
influential authors of the 20th century. Kafka strongly influenced genr
Metamorphosis"), *Der Prozess* (*The Trial*), and *Das Schloss* (*The Ca*

Betrachtung (*Contemplation*) and *Ein Landarzt* (*A Country Doctor*
ed the story collection *Ein Hungerkünstler* (*A Hunger Artist*) for
ng his novels *Der Prozess, Das Schloss* and *Amerika* (also known

Bertolt Brecht (/brɛkt/;[1][2][3] German: [ˈbɛʁtɔlt ˈbʁɛçt] (◀ listen); born
◀) *Eugen Berthold Friedrich Brecht* (help·info); 10 February 1898 – 14
was a German poet, playwright, theatre director, and Marxist.

- *The Elephant Calf* (*Das Elefantenkalb*) 1924–26/1926
- *Little Mahagonny* (*Mahagonny-Songspiel*) 1927/1927
- *The Threepenny Opera* (*Die Dreigroschenoper*) 1928/1928

Fig. 2.3 Original text sections from a larger sample of texts. The snippets contain the requested information on the authors' birth- and death date and on the titles of their works

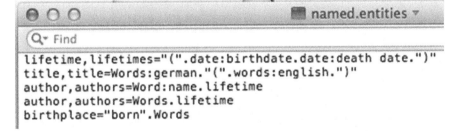

```
lifetime,lifetimes="(".date:birthdate.date:death date.")"
title,title=Words:german."(".words:english.")"
author,authors=Word:name.lifetime
author,authors=Words.lifetime
birthplace="born".Words
```

Fig. 2.4 Patterns to discover information on birth- and death dates of (German) writers and the titles of their works (original title and English translation). On the *left side* appear the labels of the patterns

In the first place, these samples may seem small to validate this ODL or to underpin its scalability. However, the inherent character of basically unstructured data distributed over different sources reflects the nature of the challenge we face

```
<author>
    <name>
        Heiner Müller
    </name>
    <lifetime>
        <birthdate>
            <date>
                1929-9-1
            </date>
        </birthdate>
        <death_date>
            <date>
                1995-30-12
            </date>
        </death_date>
    </lifetime>
</author>
```

```
<title>
    <german>
        Die Umsiedlerin
    </german>
    <english>
        The Resettler Woman
    </english>
</title>
<title>
    <german>
        Der Bau
    </german>
    <english>
        Construction Site
    </english>
</title>
```

Fig. 2.5 Example of the results rendered in XML. The discovery engine takes the patterns as defined by the users (similar to those in Fig. 2.4), detects the corresponding data in the texts (like the ones shown in Fig. 2.3), extracts these data, and stores them in XML. The tags of the XML elements (or slots) correspond to the pattern labels

in data discovery, even within the context of Big Data. The language applied in contracts and related certificates is quite uniform and not narratively complex. The document samples cover this language in its entirety and thus scale for even larger collections. In many information ecosystems, we barely have to deal with highly complex narrative forms. Due to this fact, we can consider this approach as scalable also toward thematic areas outside legal information as long as the narrative nature is relatively uniform, such as is the case for legal texts.

2.5 Conclusion

Clear vision and clear strategy for information products and services are indispensable for capitalizing on Big Data. However, what is essential for successful Big Data governance? Even though technology and modeling are substantial elements, an arms race toward more computing power and even more sophisticated models produces first of all just an impressive amount of data. If we rely too much on database performance, computer power, plug-ins to tackle new data types, and tuning of our statistical apparatus, we miss the chance to exploit the full potential of Big Data. It is, in general, difficult to satisfy consumers' needs. To say "With Big Data we produce much new information. Take what you need!" reflects a strategy doomed to failure. Of course, we will see some spectacular results. However, they only nurture the Big Data hype that is doomed to implode, in the end, like the expert system hype in the 1980s of the past century. In organizations, lofty expectations are still evoked, mainly by technology vendors promising that Big Data tools are easy to handle and showcasing other organizations that have gained significant success through Big Data. When Big Data came into vogue, the quest for improved database performance, more data mining capacities, and enriched tool support appeared almost as a knee-jerk reaction. Is successful information governance in fact the product of database power and statistical sophistication?

When we approach information governance not from a technology but from a consumer perspective, we immediately face the age-old gap between what the users expect and what IT provides. To match both sides and to produce information that meets demand, providers and consumers have to speak the same language. Information consumers want bytes and pieces that exactly meet their particular needs. They expect a certain variety of information echoing the diversity of their information demand and appreciate more meaningful information for their analytics, reporting, or whatsoever. The more Big Data meet the expectations of information consumers, the higher are the advantages. These can be manifold, far beyond business success such as growing revenues and market shares. Big Data governance should thus include semantic search and consider participatory design as the leading paradigm for implementing semantic search in Big Data analytics. It addresses also the cooperative design and management of the common vocabulary and its operationalization.

This article draws the attention to the probably most valuable resource in Big Data analytics: the information consumers. They are best not only in sketching their information requests but also in describing the information landscape they are dealing with. Of course, they sketch both things in natural language, in their individual natural language. This sketch of an information landscape takes the form of a common vocabulary. Interrelated information concepts constitute this vocabulary and reflect a semantic and abstract presentation of the information ecosystem.

A successful program for Big Data governance must include semantic search and a vigorous management of the metadata constituting the semantic layer. Common vocabularies list all concepts and terms that describe an organization's business. They are the source of the metadata that manifest the common language on one side and control information extraction and integration services on the other. We may not forget that most of Big Data is unstructured, mainly text data. This fact shifts the focus of data governance even more toward its semantic dimension.

If we take the users' information blueprints and operationalize them, that is, transform them into machine-processable instructions, we create a powerful instrument to locate required information in masses of data, in particular unstructured data. An open discovery language like the one presented here supports information consumers not only to sketch their information landscape but also to define information patterns that control the information extraction process.

Even though information consumers may manage their information discovery, they need some support from IT. Data stewards, for instance, that sufficiently know both, the domain of the organization and Big Data IT, can provide some training in information discovery. With Big Data emerged a new profession, the "chief data officer" (CDO). Even though the role of the CDO still needs to be defined, we can roughly say that the CDO understands data and their value in the context of the organization's purpose. Furthermore, this understanding is a common one, collectively shared among all information consumers within an organization. The CDO, thus, can take the role of the data steward and can help to manage the collective design of the semantic layer and its vocabularies.

There are many discovery tasks that serve individual, ad hoc, and transient purposes. Mainstream discovery, in contrast, supports reoccurring discovery requests commonly shared by large user communities and operates on large data collections, including sometimes the entire Web. We can conceive manifold scenarios for nonmainstream discovery. Users may have to analyze from time to time dozens of failure descriptions or complaints, for instance. The corresponding data collections are personal or shared among small groups and consist of bunches of PDF files or emails, for instance, barely documents on the Web. Dynamically changing small-scale requests would mean permanent system adaptation, which is too intricate and too expensive in the majority of cases. With a flexible self-service solution like ODL, information consumers can reap the benefits of automatic information discovery and sharing and avoid the drawbacks of mainstream discovery. The actual version of the ODL described here is available on sourceforge.net: http://sourceforge.net/projects/mydistiller/

References

1. Brandt DS, Uden L (2003) Insight into mental models of novice internet searchers. Commun ACM 46(7):133–136
2. Cowie J, Lehnert W (1996) Information extraction. Commun ACM 39(1):80–91
3. Ding L, Finin T, Joshi A, Pan R, Peng Y, Reddivari P (2005) Search on the semantic web. IEEE Comput 38(10):62–69
4. Fan J, Kalyanpur A, Gondek DC, Ferrucci DA (2012) Automatic knowledge extraction from documents. IBM J Res Dev 56(3.4):5:1–5:10
5. Gudivada VN, Baeza-Yates R, Raghavan VV (2015) Big data: promises and problems. IEEE Comput 48(3):20–23
6. Iwanska LM (2000) Natural language is a powerful knowledge representation system: the UNO model. In: Iwanska LM, Shapiro SC (eds) Natural language processing and knowledge representation. AAAI Press, Menlo Park, pp 7–64
7. Lohr S (2014) Google flu trends: the limits of big data. New York Times, 28 Mar 2014
8. Magaria T, Hinchey M (2013) Simplicity in IT: the power of less. IEEE Comput 46(11):23–25
9. Norman D (1987) Some observations on mental models. In: Gentner D, Stevens A (eds) Mental models. Lawrence Erlbaum, Hillsdale
10. Pentland A (2013) The data-driven society. Sci Am 309(4):64–69
11. Robertson T, Simonsen J (2012) Challenges and opportunities in contemporary participatory design. Des Issues 28(3):3–9
12. Sallam R, Tapadinhas J, Parenteau J, Yuen D, Hostmann B (2014) Magic quadrant for business intelligence and analytics platforms. http://www.gartner.com/technology/reprints.do?id=1-1QYL23J&ct=140220&st=sb. Retrieved 12 June 2014
13. Sawyer P, Rayson P, Cosh K (2005) Shallow knowledge as an aid to deep understanding in early phase requirements engineering. IEEE Trans Softw Eng 31(11):969–981
14. Tufekci Z, King B (2014) We can't trust Uber. New York Times, 7 Dec 2014
15. Viaene S (2013) Data scientists aren't domain experts. IT Prof 15(6):12–17
16. Zhao H (2007) Semantic matching. Commun ACM 50(1):45–50

Chapter 3
Multimedia Big Data: Content Analysis and Retrieval

Jer Hayes

Abstract This chapter surveys recent developments in the area of multimedia big data, the biggest big data. One core problem is how to best process this multimedia big data in an efficient and scalable way. We outline examples of the use of the MapReduce framework, including Hadoop, which has become the most common approach to a truly scalable and efficient framework for common multimedia processing tasks, e.g., content analysis and retrieval. We also examine recent developments on deep learning which has produced promising results in large-scale multimedia processing and retrieval. Overall the focus has been on empirical studies rather than the theoretical so as to highlight the most practically successful recent developments and highlight the associated caveats or lessons learned.

3.1 Introduction

Multimedia data has been called the 'biggest big data' [1]. There is an ongoing explosion in the volume and ubiquity of multimedia data. It approximately makes up 60 % of Internet traffic and 70 % of mobile phone traffic and 70 % of all available unstructured data [2]. Although in a general person's day-to-day life it may appear that multimedia data belongs to the realm of the social media, in fact there are many other sources for multimedia data. Many cities already use video cameras that gather information for security, traffic management, and safety and are increasing the numbers of these cameras and require software to quickly detect events [2], e.g., IBM Video Correlation and Analysis Suite (VCAS) [3]. In the health domain many hospitals are acquiring billions of radiological images per year that need to be analyzed correctly. The processing of satellite imagery is another source of large multimedia files. But much of the current and future multimedia content is being created by consumers.

Despite the ubiquity of images and videos that are created by consumers, it has been said that the tools for organizing and retrieving these multimedia data

J. Hayes (✉)
IBM Research, Dublin, Ireland
e-mail: HAYESJER@ie.ibm.com

M. Trovati et al. (eds.), *Big-Data Analytics and Cloud Computing*,
DOI 10.1007/978-3-319-25313-8_3

are largely still quite primitive [4] as evidenced by the lack of tools to effectively organize personal images or videos. For many consumers such information is still stored in a rudimentary folder structure with only the associated operating system providing basic indexing (and search) by time and date or filename. In this respect, we cannot discuss multimedia big data without also discussing multimedia information retrieval (MIR). Without MIR, multimedia cannot be searched effectively. The two basic functions of an MIR system are the following: (1) search – allows users to search for media content and (2) browsing and summarizing the media collection. However, this is not to suggest that professional MIR systems do not exist; they do and have a long history.

Ultimately the core challenges are how best to process multimedia big data and how to store it efficiently given the ever-increasing influx of data. In the context of MIR, a feature is any property that can be used for description and more practically can be used to aid retrieval. Semantic descriptions of the contents of an image or a video would be high-level features, whereas features that describe the color, edges, and motion are considered low level. The gap between the low-level features and the high-level features is known as the semantic gap. The description of the multimedia in terms of features has implications in terms of what processing can be done. The choices made in feature extraction or content analysis will potentially limit the types of processing, e.g., machine learning methods that can be applied. Video is an interesting big multimedia data type as it is multimodal having both visual and acoustic content. The features extracted from each modality are likely to be different, but it would make sense to combine them in some way to aid retrieval. McFee and Lanckriet highlight that in some domains, e.g., in text retrieval, there is a general acceptance as to what a good and appropriate similarity measure is, e.g., cosine similarity [5], whereas in other domains there is no obvious choice for a similarity measure. Different aspects of the data may require different similarity measures.

Multimedia big data has to be analyzed to extract information so that it can be indexed in some way and that data can be compared to other data via distance (or similarity measures). The analysis of data is (1) content based, (2) context based, or, more commonly, (3) a combination of the content and context based. The content-based approaches process the media itself to extract features, so the subsequent retrieval of multimedia will be based solely on these features. Context-based analysis uses the information associated with the data, e.g., text annotations or GPS, to extract the features that are used for search. Alternatively, there are also methods for actively annotating or assigning labels to media that should describe their content rather than just working with 'found' labels. Annotation in MIR assigns a set of labels to media that describe their content at syntactic and semantic levels. The content based methods are often used in cases where no alternative is available, i.e., there are no associated text annotations. Whatever methods are used, ultimately the querying and retrieval of multimedia data rely on being able to compare one instance with another to see how similar or dissimilar they are. The nature of the comparison is also based in the structure or data model that is used for the multimedia data, and there are several data models for multimedia data, e.g., the MPEG7 standard and it's follow-up MPEG21 [6].

However, two developments are having a major impact on the MIR: (1) as already mentioned the ever-increasing ubiquity of computing and sensing technologies, such as the emergence of social networking websites and mobile devices that are continuously creating rich media content and (2) the development of big data analytics platforms and hosts. The almost continual generation of multimedia data as well as associated sensor data has also given rise to the area of 'lifelogging' which deals with personal big data [7]. Simply put we suggest that the core problems in multimedia big data are efficiently storing, processing, indexing, and searching large volume data. Part of this is the extraction of features from the multimedia data. Search retrieval performance is a function of the feature representation and the similarity measure. Recently there has been move towards deep learning. One aspect of deep learning is that it can be thought of as a method of feature extraction, one that learns a hierarchy of features. As this feature extraction is one of the most common tasks in multimedia, then deep learning will potentially be extremely useful here. Given that a hierarchy of features is learned, it may go some way to bridging the semantic gap.

Conventional multimedia computing is often built on top of handcrafted features, which are often much restrictive in capturing complex multimedia content such as images, audios, text, and user-generated data with domain-specific knowledge. Recent progress on deep learning opens an exciting new era, placing multimedia computing on a more rigorous foundation with automatically learned representations to model the multimodal data and the cross-media interactions.

Multimedia big data information retrieval requires a scalable and efficient processing and storage framework due to the size and volume of the data. This problem requires particular architectures that can handle this, ones that are beyond the traditional database management systems. The MapReduce framework has become the de facto standard for the batch processing of huge amounts of data. In the Sect. 3.2 we report on various uses of the MapReduce framework in relation to multimedia big data and the related empirical research. We focus on the lessons that have been learned from these deployments. In Sect. 3.3 we look at deep learning which has been applied to many areas but has proven extremely useful with respect to content analysis of multimedia data. Finally in the conclusions we highlight some of the caveats that have been in both areas.

3.2 The MapReduce Framework and Multimedia Big Data

The challenge of multimedia big data is to have a scalable and efficient processing and storage framework. The MapReduce framework has become the de facto standard for the batch processing of huge amounts of data. By MapReduce framework, we include the Hadoop Distributed File System (HDFS) and other related technologies whereby data in a Hadoop cluster is held in blocks and distributed throughout the cluster. Then map and reduce functions can be executed on smaller subsets of the larger data sets.

There has been some significant recent work on the use of the MapReduce framework for the indexing, searching, and analysis of multimedia big data. Most of the examples that are outlined only use image data, although [8] uses video data. An early work in the area presented the design and implementation of an architecture that used the MapReduce framework as the Hadoop distributed file system for storage and MapReduce paradigm for processing images gathered from the web [9]. Over 1.8 billion images are uploaded and shared every day [10]. Generally, there is a wide range in the size and quality of images on the web. A standard approach with regard to storage and processing multimedia data is to apply a locality of reference whereby data is stored as close to the machine that will be undertaking the processing. This approach becomes more difficult to implement where more and more data is added and more machines are required. This may require complete redistribution of the data to place it closest to the appropriate machine(s) for processing.

3.2.1 Indexing

Using the MapReduce framework seems an obvious choice, but there is no guarantee that using this framework to index multimedia will improve performance a priori. Potentially it may constrain the possible access patterns to data and scarce resources such as RAM have to be shared [9]. Thus empirical trials are needed to provide concrete evidence of the effectiveness of this framework to this type of data. For [9] the input to the whole system is a list of image URLs and the related contextual information of the image. This input data is a tuple of {imageURL, pageurl, termvector, anchortext} that contains information on the source URLs plus the textual data in and around the image of interest. One basic question when taking an input in is does it exist already on the system? The reducer asserts whether the incoming data is new or modified or deleted based on the business logic. Regarding the tuple input, what is interesting here is that the imageURL which is unique may be associated with several pageurls and is thus referenced from multiple locations. Krishna et al. [9] found a 3:1 ratio between number of pages that reference the image and number of imageURLs. Where these are detected, the image is only downloaded once. The URL data is used to fetch the actual image from the web. A thumbnail is created from the image, and the binary image data along with the thumbnail data (and a unique key) is stored on HDFS. These are stored in separate folders with a common field so they can be joined later. Essentially, separating the data aids image-processing algorithms later so that they can easily access only image data. Separate jobs that just operate on the metadata can use this data independently. To test potential real-world uses of multimedia storage and processing system, they considered for separate use cases for real-time (sub-minute), daily, weekly, and monthly content updates. Based on this frequency of data refresh and the size of the data, the system was logically partitioned into multiple clusters in HDFS by allocating a root level folder for each cluster.

Content analysis was used to detect pornographic images with several image-processing algorithms and text analysis run as MapReduce jobs. For the experiments [9] had three setups involving (1) two nodes, (2) 100 nodes, and (3) 200 nodes. The two-node setup was used to test small feeds of about 6000 documents on the average. The 100-node setup was used to test the processing of a few million documents. The 200 node was targeted towards processing more than a billion documents. The experimental result of [9] was that there was exponential increase in time as the number of documents increased for all the three setups. The two-node setup was performed reasonably well for the real-time test case when the number of documents was less than 1000. The 100-node setup did not perform well for more than 300M documents as the processing time was close to 10 h. They observed a high initial processing time with system memory apparently being a major bottleneck. The 200 node did not show any significant improvement in time for documents in the order of few thousands when compared to the 100-node setup. However, when there were millions of documents, the 200-node cluster performed roughly six times better than the 100-node cluster. Essentially this work demonstrates that MapReduce framework is suitable for multimedia applications where there is a need to fetch, process, and store large number of documents.

In [8] their experience of adapting a high-dimensional indexing algorithm to the MapReduce model and testing it at large scale with Hadoop is outlined. They used a clustering-based high-dimensional indexing algorithm to index 30 billion SIFT (Scale Invariant Feature Transform) descriptors. These SIFT descriptors are useful as they have distinctive invariant features from images that can allow reliable matching between different views of an object or scene [sift]. They result in multiple descriptors being associated with an image; in their empirical work there are 300 SIFT descriptors per image on average [8]. Indexes can be built on these descriptors to allow fast retrieval, but it is extremely time-consuming with index over tens of millions of real-world image taking days to weeks to carry out. This is clearly an area where a scalable and efficient processing and storage framework is a necessity. When it comes to indexing, most high-dimensional indexing schemes will partition the data into smaller groups or clusters, and at retrieval time one or more of these groups are analyzed. Different techniques use different procedures to create the partitioned groups. They chose to use eCP (extended cluster pruning) to create clusters that group the descriptors and in which query vectors are matched against. This clustering approach is related to the k-means approach but is more I/O oriented as it assumes the data will be too large to fit in memory and will reside on secondary storage. When the data set is large, the clusters are organized into a hierarchical structure to speed up access. Using the MapReduce framework, they outline how to build an index and how to search it and report the results from large-scale experiments.

3.2.2 Caveats on Indexing

Ultimately, [8] shows how the MapReduce framework can be applied to indexing algorithms and in turn how scalable they are. A large number of experiments were carried out, and a number of interesting caveats or lessons are put forward. Effectively they evaluated index creation and search using an image collection containing roughly 100 million images with about 30 billion SIFT descriptors. The first caveat is that the performance of indexing the data was hurt when the data collection occupied many blocks and when each map task needed to load a lot of auxiliary information at start-up time [8]. Data which occupies many blocks requires many map tasks to be run in the MapReduce framework. One way to avoid this is to increase the size of the blocks of data to a value beyond that recommended by Hadoop. This has to be balanced against having nodes run out of disk space as the temporary area buffering the data produced by mappers and consumed by reducers is used up more quickly when blocks are big.

Another caveat that may hold for many different processing tasks is that map tasks are independent of each other and each requires their own individual auxiliary information at start-up. When this auxiliary information is large, then each map task consumes a significant portion of the RAM available on a node. This in turn can result in map tasks that are unable to run inside every available core in a node because there is not enough RAM. In the context of [8] and the indexing algorithm, each map task required information on the tree structure required for building the index. This data could not shared between map tasks running on the same node due to the architecture of Hadoop.

3.2.3 Multiple Multimedia Processing

Clearly as we have mentioned earlier, one aspect of big data is the variety of the data. Yet the approaches that we have described above largely focus on image analysis. Of course there is a large degree of variety with this image data especially where it is mined from the web; however it is just one media type, and generally only a type of processing has been applied. What if we wished to use more than one processing type? One possibility is to use the MapReduce framework to handle distinct multiple multimedia processing techniques simultaneously.

Chen et al. [11] designed a parallel processing mechanism for multiple multimedia processing techniques based on the existing MapReduce framework for GPUs. Programming interfaces for GPUs are generally designed for graphics applications, and 'Mars' is a MapReduce framework that hides the programming complexity of the GPU and allows programmers to use a more familiar MapReduce interface [12].

Targeting the MapReduce framework at GPUs rather than CPUs has advantages as there is a higher bandwidth memory access rate with CPU. This results in a more efficient performance for the MapReduce framework. More specifically [mrChecn] focused on the deployment of more than one multimedia processing program on the GPU MapReduce framework and how best to divide work units to create an efficient parallelization. A middleware manager is implemented in the Mars framework, and this is responsible for managing all the service requests of the multimedia processing programs. In fact this middleware manager replaces the Mars Scheduler in the original Mars framework. Their results suggested that the average processing speed under the proposed mechanism is improved by 1.3 times over the previous framework. Continuing in the vein of using the MapReduce framework to carry out multiple tasks [13] explore the framework for large-scale multimedia data mining. They present empirical studies on image classification, video event detection, and near-duplicate video retrieval. They used a five-node Hadoop cluster to demonstrate the efficiency of the proposed MapReduce framework for large-scale multimedia data mining applications. The MapReduce framework can read in input data in several different formats. However, many traditional Hadoop implementations are targeted as processing text data and thus may face difficulties in dealing with multimedia data types such as video and images [13]. To work around this problem, one 'mapper' was used to process an entire input file by customizing to two classes in the standard Hadoop implementation (InputFormat and RecordReader). This resulted in image and video files being presented in a format that could be processed directly.

Processing of the multimedia content for feature extraction is done by harnessing third-party toolkits. During the MapReduce phase for each multimedia file passed to a corresponding computer node in the form of byte stream as an entire load, each mapper receives the data and generates a new copy of the file in the local disk [13]. At this point, executable programs are invoked and run natively on the process the temporary file and return the extracted features. This process occurs in parallel with other nodes and does not interfere with them. Ultimately the extracted features are returned as part of key/value pair with the name of the file as the key and extracted features as the value. In the Reduce phase, all these key/value pairs are aggregated together to output a final image/video feature data set [13]. The multimedia was also processed for clustering (k-means clustering), generating a 'bag of features' of image or a video as an orderless collection of keypoint features, image classification, and video event detection and near duplicate video retrieval. Only experimental results from three kinds of applications are listed: image classification and video event detection and near-duplicate video retrieval. However, these are common multimedia processing tasks.

Extensive experimental results on several benchmark image/video data sets demonstrate that their approach can efficiently reduce the processing time for three applications examined. Three scenarios were used to show the scalability of the approach: (1) only one node was used, (2) a three-node cluster was used, and (3) five computers are used as a five-node cluster. For the image classification application; processing time significantly decreased across each scenario as it did for the video event detection tasks. Overall, the performance was consistent with the baselines for the tasks. Finally for near-duplicate video retrieval, again a decrease in processing time across the scenarios was found.

3.2.4   Additional Work Required?

If we briefly compare the approaches listed so far we can see that [13] modified the standard Hadoop implementation to read in the multimedia files more easily. They then used the MapReduce framework to distribute files in such a way that third-party toolkits were run natively to extract results for the Reduce part of the processing. By doing so, they could leverage already existing external tools and did not have to modify algorithms/write code to take a MapReduce form. This approach may prove useful to research teams who use third-party toolkits to process multimedia data. But it still required modification of the standard MapReduce framework.

Also [9] found that they had to port the eCP algorithm to fit into the MapReduce framework. The original algorithm creates clusters but uses information on disk I/O, e.g., the cluster count is generally chosen such that the average cluster size will fit in on one disk, as part of the process. Thus it would appear that any algorithm that takes account of disk I/O when implemented in a serial fashion will require additional work to fit into the MapReduce framework. They also highlighted a number of problems with using the MapReduce framework 'out of the box'. Multimedia files may be larger than the normal Hadoop block sizes. Clearly this too was a problem for [13], and this resulted in changing how files were read in their deployment. So in short, although the MapReduce framework has proven useful, it appears that additional work is required to work with multimedia data.

Thus far we have outlined recent results using the MapReduce framework. In contrast and also potentially in addition, Apache Storm can be viewed as an in-memory parallelization framework. Mera et al. [14] examines empirically the choice of Storm versus the MapReduce framework for the task of extracting features for indexing. In order to test different workloads on both the MapReduce and the Storm frameworks, three workloads were created: (1) one with ten thousand images; (2) one with a hundred thousand images, and (3) one with all one million of images. Overall it was found that from the scalability point of view, the Storm approach has a better scalability in most cases [storm]. But the MapReduce framework scaled better than Storm when more resources and data are available. With the caveat that the direct comparisons of the MapReduce framework with the Storm framework will require further empirical studies using larger data and more resources, [14] hypothesize that Storm scales better in small infrastructures, while the MapReduce framework scales better in larger infrastructures. In effect as algorithms and techniques from computer vision, image processing, and machine learning are often the basis for multimedia data mining, then general research from these area is also relevant. We explore an important technique for machine learning and multimedia data in the next section.

3.3   Deep Learning and Multimedia Data

Deep learning is a family of machine learning algorithms that attempt to model high-level abstractions in data by employing deep architectures composed of multiple levels of representation [15]. It is 'deep' in contrast to other machine learning algorithms, e.g., support vector machines, as it is composed of multiple layers of adaptive nonlinear transformations. It can be said that deep learning attempts to learn distinct levels of representation that mimics the human brain. Indeed, it has been recently shown that deep neural networks match the ability of the primate brain to recognize objects when shown them rapidly (100ms) [16]. The human brain has a deep architecture as it processes information through multiple stages of transformation and representation, e.g., in human speech production. In a similar way with respect to multiple levels of transformation and representation, deep learning creates multiple levels of features or representations of the data, starting with the breakthrough of [17] which introduced a deep belief network (DBN) composed of stack of restricted Boltzmann machines (RBMs).

One taxonomy is to make distinctions between (1) generative deep architectures, (2) discriminative deep architectures, and (3) hybrid deep architectures [15]. In the first category are architectures that are generally used for unsupervised feature learning and are intended to characterize the high-order correlation properties of the observed data [15]. The discriminative deep architectures are intended for pattern classification. The final category is those models which are intended to discriminate but use the outcomes of the generative architectures. Deep learning has had a massive impact as it has worked successfully in several different domains and on occasion has made a massive leap over the then state of the art [18]. Indeed, [18] is probably the most cited deep learning paper in the literature and involved the training of a large, deep convolutional neural network to classify the 1.3 million high-resolution images in the LSVRC-2010 ImageNet training set into 1000 different classes. They achieved top-1 and top-5 error rates of 39.7 % and 18.9 % which were considerably better than the previous state-of-the-art results. In the area of speech recognition, it has also achieved state-of-the-art performance in several other areas [19]. This along with the adoption of deep learning by various companies such as Google, Apple, and IBM has shifted much of the research direction in multimedia big data (and in many other fields) to deep learning.

There are a large number of different approaches to deep learning with two of the most common being deep belief networks and convolutional neural networks (CNN) [20]. Deep learning using a convolution structure has been applied very effectively in computer vision and image recognition, and CNN are the most common with respect to image processing and especially object recognition. Convolution can be interpreted in this context as a filter. Many convolution filters exist and are common effects of image-processing applications, e.g., blur, emboss, sharpen, find edges, and so on. However, multimedia is not just solely image data. A deep neural network that uses convolution structure is neural network where each convolutional layer performs a 2D convolution of its input maps with a square filter [21]. In practice convolutional neural network is composed of alternating layers of convolution and pooling [22]. The aim of 2D convolution is to extract patterns found within subparts of the input images that are common throughout the data set.

In practice multimedia data is often multimodal, e.g., web-based images are often associated with text; videos contain both visual and audio data; text data and image data on social media sites can be related to social networks. Note that most of the examples listed in the previous section were targeted at a single mode. This makes sense at one level as the different modes of data often have very distinct representations, e.g., text is discrete and is generally represented by very large and sparse vectors in an information retrieval setting where images are often represented by dense tensors that exhibit strong local correlations [23]. This makes processing and learning from multimodal data technically challenging as there will be different kinds of representation and different associated statistics.

We will briefly review the impact of deep learning on multimedia big data before considering how it is being applied to multimodal data which is still a new frontier with respect to this area. From the viewpoint of content-based image retrieval (CBIR), [23] pose the question of how much improvements in CBIR tasks can be achieved by exploring the state-of-the-art deep learning techniques for learning feature representations and similarity measures and offer some extensive empirical results onto this question. In this context 'content based' means that the image retrieval aspect is based on the contents of the image rather than the metadata associated with the image such as keywords, tags, etc. As mentioned earlier, there is a semantic gap between the content-based features extracted and the types of 'concept-based' queries that users often give. Wan et al. [23] applied convolutional neural networks (CNNs) to learning feature representations from image data. They then conducted an extensive set of empirical studies based on typical CBIR tasks. The implementation of the CNN was based on that of [18] and adapted their publicly released C++ implementation [24]. The data set they chose was the LSVRC-2010 ImageNet training set. Briefly, this consists of large hand-labeled ImageNet data set (10,000,000 labeled images depicting 10,000+ object categories). The general task associated with this data set is to build a model and then test on images with annotation, i.e., no labels or segmentation data, to produce labels that specify what objects are present in the test images. This is a typical classification task and does not touch on some work that is needed to build a retrieval system.

After building the CNN [23], they took the activations of the last three fully connected layers as the feature representations. The assumption is that features from the lower levels are less likely to have rich semantic content. However, these features may not be useful for information retrieval on their own. So as well as just using these extracted features [23] also examined two additional scenarios. In one a metric learning technique was used to refine the feature representation of the extracted features. In the final scenario they retrained a deep CNN model with classification or similarity loss function which is initialized by the original ImageNet-trained model. Wan et al. [23] report that features extracted directly from the CNN can be used effectively for the information retrieval tasks. Although trained on the ImageNet data set, several other data sets were used for a variety of retrieval tasks.

For one data set (Caltech256) there were promising results for the deep learning features. This indicated that the model was capturing high semantic information. The deep learning feature representations consistently outperformed conventional handcrafted features across all data sets. It was found that further similarity learning could improve the retrieval performance of the initial deep learning features when the model was applied to a new domain. Additionally retraining the model also improved retrieval performance better than the improvements made by similarity learning [23]. These initial results are positive and indicate the efficacy of this research direction for multimedia information retrieval.

Similarly deep learning was also used for feature extraction in [25] as input to other machine learning techniques. The application area was image retrieval and more specifically the MSR-Bing Image Retrieval Challenge [26]. The traditional approach to web-based image retrieval is to use text information and thus treat retrieval using the tools of web-based information retrieval which uses text as the source of its data objects. The text information is generally the text that is associated, in close proximity, to the image on web page or in a document. However, in reality this text may not actually contain information that perfectly describes the image, i.e., the objects that are in it, and in some cases there may not be text at all. Hua et al. [25] used the output of CNN models to be a concept-level representation of an image. They trained two CNN models to return the category probability vector of an input image against the categories of the training data set. There were 1119 categories selected from the MSR-Bing Image Retrieval Challenge training data set, and one model used probabilities over the 1119 categories output from CNN as the representation of the image. Another model was built using the ImageNet data set and had 1000 categories. Between the two, they obtained 1119 word categories with around 680,000 images for CNN training. They argue that this representation helps to bridge the semantic gap as it contains more semantic information than the standard low-level features alone.

With respect to 3D object recognition, [27] used a deep belief net where the top layer model used a third-order Boltzmann machine. This model was applied to a data set that contained stereo-pair images of objects under different lighting conditions and viewpoints (the NORB database). The first version of the model had an error rate close to the best performance on this task, 6.5 % versus 5.9 %. This was much better error than just using an SVM which was reported to have an error rate of 11.6 %. By taking the same amount of labeled data but this time adding extra unlabeled data, the model achieved an error rate better than the previous best with error rate improving from 6.5 to 5.2 %.

The health domain is increasingly generating important multimedia data used to inform the decisions of clinicians. In [21] a deep neural network was built to classify related stacked images from the ISBI 2012 EM Segmentation Challenge [28]. The data represented two portions of the ventral nerve cord of a Drosophila larva. For the testing, two manual segmentations by different expert neuroanatomists were obtained. The results were that the model outperformed other competing techniques by a large margin and in one of the metrics the approach outperformed a second human observer.

Zeiler et al. developed a new model that is based on the standard convolutional neural network architecture but changed the pooling scheme [22]. The stochastic pooling scheme they propose can be used in any standard convolutional neural network architecture. Their pooling scheme is made by sampling from a multinomial distribution formed from the activations. The resulting model was applied to four different image data sets: MNIST, CIFAR-10, CIFAR-100, and Street View House Numbers (SVHN). For the experiments on the CIFAR-10 data set, three models were trained: one using the novel stochastic pooling, one using average pooling, and one using max pooling (which is the standard). The CIFAR-10 data set is made up of 10 classes of natural images with 50,000 training examples in total (with 5,000 per class). Their model with the stochastic pooling had the lowest error rate of 15.13, while average pooling and max pooling had rates of 19.24 and 19.40, respectively. The model using the stochastic pooling outperformed the other models across the rest of the data sets too.

Multimodal learning involves relating information from multiple multimedia sources. In [29] learning representations for speech audio which are coupled with videos of the lips are the focus. These videos of lips are considered 'mid-level' as opposed to the low level of raw pixels. This learning process follows the standard feature learning, supervised learning, and testing phases. The integration of the various modes can be done in different ways, and [29] consider three: multimodal fusion, cross modality learning, and shared representation learning. In multimodal fusion the data from all modalities is available through all the phases. In cross modality learning, only during feature learning is data from all modalities made available, and during the remaining supervised training and testing phase only data from a single modality (the one of interest) is provided. So we may want to learn features from video for the learning and testing phases, but we would use the video and audio to extract features in the first phase. Finally in the shared representation approach, the different modalities are presented for supervised training and testing.

The baseline that is used in [29] is to train a deep learning network using RBM model for audio and video separately. These are to create a new representation for the data. Ultimately the results show that there is an improvement by learning video features with both video and audio compared to learning features with only video data. It should be noted that the video in these examples was of lip readings. This has features that more clearly relate to the audio than say a general YouTube video. However, this was the target of the study. The data sets used have human subjects repeating letters or digits, and [29] demonstrated the best published visual speech classification on one of these data sets.

Given the focus of the previous section was on scalability, we should also ask well if deep learning models do scale. Coates et al. used a cluster of GPU servers with InfiniBand interconnects and MPI to train 1 billion parameter networks on just three machines in a couple of days [30]. After testing the system, they trained a much larger network from the same data set. This network with 11 billion parameters was created using just 16 machines. One of the caveats in [30] is that new algorithms or analysis methods may potentially be needed for the very large sizes of networks, very large compared to what has typically been processed before.

3.4   Conclusions

This chapter gave an overview of the recent developments in the area of multimedia big data, the biggest big data, with a focus on empirical studies rather than the theoretical so as to highlight the most successful and useful.

The use of the MapReduce framework is becoming increasingly popular in the area of multimedia big data. It can effectively provide a scalable framework as evidenced in [8, 9, 11]. However, it appears that to process some multimedia data types modifications to the MapReduce framework must be made first. Perhaps over time some of these modifications will become standard in the use of multimedia in this framework. But currently in implementing the MapReduce framework, a researcher will likely have to adapt how certain multimedia files are read in. However, this additional work can be balanced against how the framework was used to run third-party toolkits natively [11].

Deep learning has had and is continuing to have a large impact in several areas including in the processing of multimedia data. As it can work with either unlabeled data or weakly labeled data to analyze the content of multimedia, it has allowed researchers to work with larger and larger data sets as they do not have to rely on human generated/handcrafted features. Many of the most cited papers relate to image processing and classification which can be viewed as content analysis when one considers that contents of the multimedia have been processed and features extracted. Although larger and larger data sets are being modeled by deep learning networks, one of the caveats in [30] is that new algorithms or analysis methods may be needed for the very large sizes of networks, but further empirical research is needed to explore this.

References

1. (2015) Special issue on multimedia: the biggest big data. IEEE Trans Multimed 17(1):144
2. Smith JR (2013) Riding the multimedia big data wave. In: Proceedings of the 36th International ACM SIGIR Conference on Research and Development in Information Retrieval (SIGIR '13), Dublin, 28 July–01 Aug, pp 1–2
3. Video correlation and analysis suite from IBM. http://www.ibm.com/smarterplanet/ie/en/smarter_cities/solutions/solution/A863656S12083Q41.html
4. Chang EY (2011) Foundations of large-scale multimedia information management and retrieval: mathematics of perception. Springer, New York
5. McFee B, Lanckriet G (2011) Learning multi-modal similarity. J Mach Learn Res 12:491–523
6. Burnett I, Van de Walle R, Hill K, Bormans J, Pereira F (2003) MPEG-21: goals and achievements. IEEE MultiMed 10(4):60–70
7. Gurrin C, Smeaton AF, Doherty AR (2014) LifeLogging: personal big data. Found Trends Inf Retr 8(1):1–125
8. Moise D, Shestakov D, Gudmundsson G, Amsaleg L (2013) Indexing and searching 100M images with map-reduce. In: Proceedings of the 3rd ACM Conference on International Conference on Multimedia Retrieval (ICMR '13), Dallas, 16–19 Apr, pp 17–24

9. Krishna M, Kannan B, Ramani A, Sathish SJ (2010) Implementation and performance evaluation of a hybrid distributed system for storing and processing images from the web. In: 2010 IEEE Second International Conference on Cloud Computing Technology and Science (CloudCom), Indianapolis, 30 Nov-03 Dec, pp 762–767

10. Meeker M (2014) Internet Trends 2014 – Code Conference

11. Chen SY, Lai CF, Hwang RH, Chao HC, Huang YM (2014) A multimedia parallel processing approach on GPU MapReduce framework. In: Proceedings of the 7th International Conference on Ubi-Media Computing and Workshops (UMEDIA), Ulaanbaatar, 12–14 July, pp 154–159

12. He B, Fang W, Luo Q, Govindaraju NK, Wang T (2008) Mars: a MapReduce framework on graphics processors. In: Proceedings of the 17th International Conference on Parallel Architectures and Compilation Techniques (PACT'08), Toronto, 25–29 Oct, pp 260–269

13. Wang H, Shen Y, Wang L, Zhufeng K, Wang W, Cheng C (2012) Large-scale multimedia data mining using MapReduce framework. In: IEEE 4th International Conference on Cloud Computing Technology and Science (CloudCom'12), Taipei, 3–6 Dec, pp 287–292

14. Mera D, Batko M, Zezula P (2014) Towards fast multimedia feature extraction: Hadoop or storm. In: IEEE International Symposium on Multimedia (ISM'14), Taichung, 10–12 Dec, pp 106–109

15. Deng L (2014) A tutorial survey of architectures, algorithms, and applications for deep learning. APSIPA Trans Signal Inf Process 3:e2

16. Cadieu CF, Hong H, Yamins DLK, Pinto N, Ardila D et al (2014) Deep neural networks rival the representation of primate IT cortex for core visual object recognition. PLoS Comput Biol 10(12):e1003963

17. Hinton GE, Osindero S, Teh Y (2006) A fast learning algorithm for deep belief nets. Neural Comput 18:1527–1554

18. Krizhevsky A, Sutskever I, Hinton GE (2012) ImageNet classification with deep convolutional neural networks. In: Proceedings of the Advances in Neural Information Processing Systems (NIPS'12), Lake Tahoe, Nevada

19. Hinton G, Deng L, Yu D, Dahl GE, Mohamed A, Jaitly N, Senior A, Vanhoucke V, Nguyen P, Sainath TN, Kingsbury B (2012) Deep neural networks for acoustic modeling in speech recognition: the shared views of four research groups. IEEE Signal Process Mag 29(6):82–97

20. Chen X-W, Lin X (2014) Big data deep learning: challenges and perspectives. IEEE Access 2:514–525

21. Ciresan D, Giusti A, Gambardella L, Schidhuber J (2012) Deep neural networks segment neuronal membranes in electron microscopy images. In: Proceedings of the Advances in Neural Information Processing Systems (NIPS'12), Lake Tahoe, 03–08 Dec, pp 2852–2860

22. Zeiler M, Fergus R (2013) Stochastic pooling for regularization of deep convolutional neural networks. CoRR, abs/1301.3557

23. Wan J, Wang D, Hoi SCH, Wu P, Zhu J, Zhang Y, Li J (2014) Deep learning for content-based image retrieval: a comprehensive study. In: Proceedings of the ACM international conference on multimedia (MM'14), Orlando. ACM, New York, pp 157–166

24. High-performance C++/CUDA implementation of convolutional neural networks. https://code.google.com/p/cuda-convnet/

25. Hua J, Shao J, Tian H, Zhao Z, Su F, Cai A (2014) An output aggregation system for large scale cross-modal retrieval. In: IEEE International Conference on Multimedia and Expo Workshops (ICMEW'14), Chengdu, 14–18 July 2014, pp 1–6

26. MSR-Bing Image Retrieval Challenge. http://research.microsoft.com/en-US/projects/irc/acmmm2014.aspx

27. Nair V, Hinton G (2009) 3-D object recognition with deep belief nets. In: Proceedings of the Advances in Neural Information Processing Systems (NIPS'12), Lake Tahoe, 03–08 Dec, pp 1339–1347

28. ISBI 2012 EM Segmentation Challenge. http://brainiac2.mit.edu/isbi_challenge/

29. Ngiam J, Khosla A, Kim M, Nam J, Lee H, Ng A (2011) Multimodal deep learning. In: Proceedings of the 28th International Conference on Machine Learning (ICML11), Bellevue, USA, 28 June-02 July, pp 689–696
30. Coates A, Huval B, Wang T, Wu D, Ng A, Catanzaro B (2013) Deep learning with COTS HPC systems. In: Proceedings of the 30th International Conference on Machine Learning (ICML13), Atlanta, 16–21 June, pp 1337–1345

Chapter 4
An Overview of Some Theoretical Topological Aspects of Big Data

Marcello Trovati

Abstract The growth of Big Data has expanded the traditional data science approaches to address the multiple challenges associated with this field. Furthermore, the wealth of data available from a wide range of sources has fundamentally changed the requirements for theoretical methods to provide insight into this field. In this chapter, a general overview on some theoretical aspects related to Big Data is discussed.

4.1 Introduction

Big Data has been attracting increasing attention from research communities, which has contributed to a remarkable advance in the theories and applications to address the multitude of challenges posed by this field. However, the very nature of Big Data is based on a variety of multidisciplinary topics, which make an integrated approach to Big Data a challenging task.

Big Data is characterised by the *4Vs*, namely, *volume, velocity, variety* and *veracity* [1].

In fact, data is created at an exponential pace, in all forms and shapes from potentially a limitless number of sources. Therefore, the management, assessment, verification and extraction of such wealth of information pose a variety of challenges. The accuracy in these tasks may lead to more confident decision-making and, in turn, greater operational efficiency, cost reductions and reduced risk.

As a consequence, any system able to address such challenges must be able to integrate several techniques and methods from a variety of multidisciplinary scientific fields. However, such integration is often very difficult, if not impossible at times, since different tools have different theoretical backgrounds, which might be incompatible.

Therefore, there is an increasing need for a unified set of theories, which can address the complexity of Big Data.

M. Trovati (✉)
Department of Computing and Mathematics, University of Derby, Derby, UK
e-mail: M.Trovati@derby.ac.uk

© Springer International Publishing Switzerland 2015
M. Trovati et al. (eds.), *Big-Data Analytics and Cloud Computing*,
DOI 10.1007/978-3-319-25313-8_4

In this chapter, we will give an overview of some theoretical approaches based on the topological properties of data, which can be successfully exploited to provide an elegant yet powerful set of tools to advance our current understanding of data science.

4.2   Representation of Data

Topology theory has a long history, and it is believed to have been first introduced by Euler in 1736. In his influential paper "*Solutio problematis ad geometriam situs pertinentis*", Euler discussed the solution of the Königsberg bridge problem, in which the concept of distance was not at its core [2].

Subsequently, the concept of dimension has been generalised by mathematicians over the last few centuries to develop and make it central to its theoretical foundation, the idea of topological invariants [2].

Data tends to have many components, depending on its size, nature and, ultimately, what features are capturing. Therefore, data is typically described in multidimensional vector spaces, where each component refers to a specific feature. However, the potential high number of dimensions is likely to negatively impact user interaction, since this tends to be difficult to be visualised. Therefore, the ability to understand the *geometry* of such data is crucial in obtaining a full picture of what it refers to, as well as at its full meaning. One of the most basic geometric aspects, which is investigated, is clustering. Clustering is a conceptually simple instance of machine learning, where the geometric understanding of the dataset allows determining whether specific patterns appear.

As a simple example, consider a digital image. This consists of a set of pixels, which are simply dots with a certain colour feature. The challenge is to understand what image all these pixels may refer to. For example, if it depicts a human face, we would need to identify the different parts, such as eyes, ears, etc. In this case, clustering would enable us to classify and identify the groups of pixels assigned to those areas. However, depending on the size and accuracy of such sets of pixels, suitable methods are required to provide scalable, precise and efficient solutions.

There are a variety of mathematical techniques to address such challenge, and one that is rapidly gaining increasing popularity is topology.

For a huge variety of complex datasets, it is very hard, if not impossible, to determine a feasible number of hypotheses to capture the relevant information from such dataset. As a consequence, the ability to explore and understand data, without a prior formulation of any hypothesis, is crucially important.

In fact, as discussed in [3], most of the current approaches are based on the identification of specific relationships as one of the first steps.

Topology aims to understand the "shape" of data as a high-dimensional geometrical entity, in order to extract meaningful information.

The approach based on a cross-contamination between topology and data science has created a new research field called topological data analysis (TDA) [4]. Its

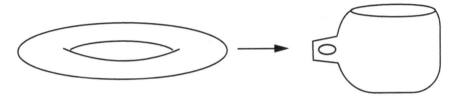

Fig. 4.1 The deformation of a doughnut into a coffee mug

focus is shape recognition within data, which can successfully lead to the full identification of features in the data.

Loosely speaking, topology classifies objects based on their invariant properties. A well-know example is the similarity between a doughnut and a coffee mug. If we were to "deform" a doughnut as depicted in Fig. 4.1, we would obtain a coffee mug.

Topology is often defined on the metric of the space that it has to be applied to. As a consequence, there is no direct need to restrict to a specific coordinate system, as opposed to many other data mining techniques. This allows a more flexible approach, which can be easily extended to a variety of contexts and platforms.

Furthermore, topology investigates objects which are invariant under specific deformations, as long as they do not change a shape "too much". This trivial and common property in data analysis is probably better understood by considering a simple example. Whenever we produce a handwritten text, our aim is to create a set of legible sentences, which can be understood by everybody else. In particular, we expect that any individual can identify the meaning associated to each handwritten word, even though my handwriting style is likely to appear differently from anyone else's.

However, such differences need to be consistent and follow specific rules, so that, for example, an "O" has only one clear loop. It could be slightly stretched, squashed or slanted, but it still would be possible to identify it as an "O". In particular, the topological invariance property implies that a circle, an ellipse and the boundary of a pentagon are all topologically similar as they all have a "hole" so that if I stretch any of them, I could still get any of the others [3].

The main advantage of such property is that it does not matter how good my "O" looks, as long as it has the general property of a loop, so that it allows it to be far less sensitive to noise. In fact, one of the objectives of topology in general is the emphasis on global properties, rather than on local ones.

In fact, one of the objectives of algebraic topology in general is the emphasis on global properties, rather than on local ones.

Another crucial aspect of topology, or algebraic topology in this particular case, is that shapes are approximated by appropriate triangulations which identify the corresponding shape using a simplicial complex or a network. When representing the shape of the physical boundaries of a country, for example, we approximate it by a well-defined segmentation, or a 1-dimensional set of shapes, which still convey sufficient accurate information to describe the corresponding country.

4.3 Homology Theory

The fundamental starting point of TDA is the definition and identification of appropriate homology groups [4].

Homology groups are algebraic entities, which quantify specific topological properties in a space. Although it does not capture all topological aspects of a space as two spaces with the same homology groups may not be topologically equivalent, two spaces that are topologically equivalent must have isomorphic homology groups. Loosely speaking, homology formalises the properties of groups that are relevant according to specific rules. Furthermore, an important aspect of homology theory is that it provides a theoretical description which can be expressed via computationally efficient techniques, with a variety of data science application.

The main motivation of homology is the identification of objects that follow specific invariant rules, very much like in the above example. More specifically, assume we have a set of data represented by points in an n-dimensional vector space. How can we extract information in a meaningful way, whilst ensuring both efficiency and accuracy?

Consider, for example, a ball of radius r around each point in the dataset. If r is too small, each ball would only contain a single point, and we would only have a topology consisting of disjoint balls. On the other hand, r is too large, and the induced topology is just one big ball containing all the point. Therefore, we need to have a radius to avoid such extreme cases and obtain enough information to capture some of the geometric structure of the underlying geometric object. Furthermore, in every dataset, there is always some level of noise, which should be addressed and perhaps ignored in order to only extract relevant information. Therefore, when choosing the "right" radius, it is important to define balls that capture "good" information. Whatever properties associated with the corresponding topology, the power of such an approach is that the topology reveals geometric features of the data set that is independent of how it is represented in lower dimensions, whilst minimising the impact of noise.

4.3.1 Simplicial Complexes

The fundamental components of homology are *simplicial complexes*, which consist of collections of simplicials [3]. These are based on specific of a space. This chapter will not discuss triangulations in details, and for a detailed description, please refer to [3]. In a nutshell, triangulation is the process of covering a shape with joined, nonoverlapping polyhedra, which can be viewed as a "shape approximation".

Strictly speaking, a triangulation does not only refer to 2-dimensional objects, and in fact, they are in general defined as polyhedral.

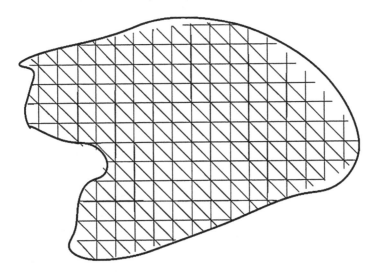

Fig. 4.2 An example of triangulation and mesh

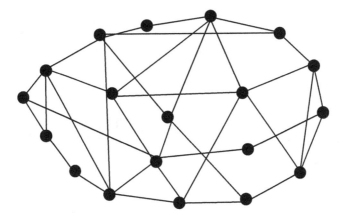

Fig. 4.3 A convex hull

Figure 4.2 depicts an example of triangulation, where it can be easily seen that such process allows an effective approximation of the corresponding shape. Clearly, such approximation depends on how all the polyhedra are defined.

The triangulation of a space S is defined by a convex combination of the points in S. This can be written as an affine combination where all the weights are non-negative or, in other words, i.e. $w_i \geq 0$ for all i.

The convex hull of S, denoted conv S, is the set of all convex combinations of points in S, as shown in Fig. 4.3.

Tetrahedra, triangles, edges and vertices are all instances of simplices as triangulation refers to any dimension. For example, a 0-simplex is a vertex, a 1-simplex is an edge, and a 2-simplex is a triangle, as depicted in Fig. 4.4.

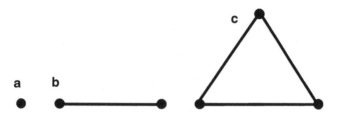

Fig. 4.4 A depiction of a 0-dimensional (**a**), 1-dimensional (**b**) and 2-dimensional (**c**) simplicials

Recall that a point set C is convex if for every pair of points a, b belonging to C and the line segment joining a and b is included in C [3].

Simplices and convex polyhedral, in particular, are convex hulls of finite point sets, where k-simplices are the "simplest" possible k-dimensional polyhedra.

A simplicial complex is defined as a collection of simplicial, such that they contain every face of every simplex in it and that the intersection of any two of its simplicials is either empty or it is a face belonging to both of them.

In particular, a k-simplex is said to have dimension k. A face of s is a simplex that is the convex hull of a non-empty subset of P. Faces of s may have different dimensions from zero, i.e. vertices, to k, as s is itself a face of s. Furthermore, the $(k-1)$-faces of s are called facets of s, so that s has $k+1$ facets. For instance, the facets of a tetrahedron are its four triangular faces.

4.3.2 Voronoi Diagrams and Delaunay Triangulations

A specific example of a triangulation is Voronoi diagrams and Delaunay triangulations [3]. These are based on the concept of distance between points in a space. In particular, the notion of neighbourhood plays a crucial role. This is, as the word suggests, the set of points within a certain distance a specific point.

Voronoi diagrams and Delaunay triangulations provide a method to approximate shapes based on the concept of neighbourhood in the discrete domain. More specifically, a Voronoi diagram consists of a collection of cells, or Voronoi cells. These are defined as the sets of points V_x so that no other point is closer to it than x and each Voronoi cell is a convex polygon. Delaunay triangulations follow Voronoi diagrams as they are defined by joining the centres of each Voronoi cell, as depicted in Fig. 4.5

Even though Fig. 4.5 refers to 2-dimensional Voronoi diagrams, the same definition applies to any arbitrary dimension. In such case, a Delaunay triangulation consists of polyhedra.

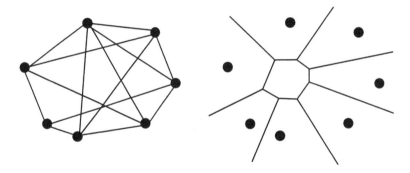

Fig. 4.5 A Voronoi and Delaunay triangulation

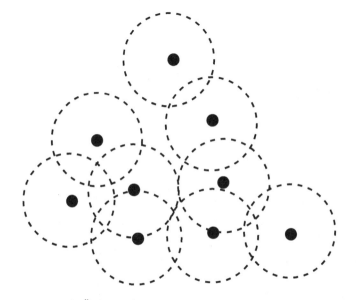

Fig. 4.6 An example of a Čech complex

4.3.3   Vietoris and Čech Complexes

Another important example of triangulation includes the Vietoris complex, which is a simplicial complex based on a distance d by forming a simplex for every finite set of points that has diameter at most d.

In other words, it has the property that the if distance between every pair of points is at most d, then this will define a complex.

A Čech complex, on the other hand, is defined by a set of balls with a specific radius. Points are in the same cell if their corresponding balls have non-empty intersection, as depicted in Fig. 4.6.

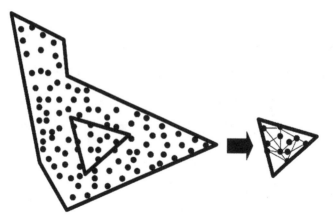

Fig. 4.7 An example of a graph-induced complex

The main difference with a Vietoris complex is that it only considers balls which have pairwise intersections.

4.3.4 Graph-Induced Complexes

Although the formulation of a Vietoris complex tends to be simple to compute and provides an efficient tool for extracting topology of sampled spaces, its size tends to be very large. In [5], the graph-induced complex is introduced. This approach provides an improvement as it works on a subsample but still retains the descriptive power of capturing the topology as the Vietoris complex. The main advantage of this approach is that it only requires a graph connecting the original sample points from which it defines a complex on the subsample. The consequence is that the overall performance is much more efficient (Fig. 4.7).

4.3.5 Chains

An important concept in homology is a *chain*. More specifically, a p-chain c in K is a formal sum of p-simplices added with some coefficients, that is, $\sum a_i s_i$, where s_i are the p-simplices and a_i are the coefficients. In particular, if $s = \{v_0, \ldots, v_p\}$, we define the boundary as $D_p s = \sum \hat{a}_i$, that is, we omit the p-th element of s.

Extending D_p to a p-chain, we obtain a $(p-1)$-chain, that is, $D_p : C_p \to C_{p-1}$.

The property that $(p-1)$-chains exhibit is that

$$C_p \to C_{p-1} \to C_{p-2} \cdots \to C_0 \to C_{-1} \to 0.$$

Another useful concept in algebraic topology is *Betti numbers*. Informally, the Betti number indicates the maximum number of cuts that can be carried out so that a surface is not divided into two separate pieces [6].

The combination of the theoretical approaches discussed above can provide a powerful set of tools to identify the relevant information from datasets, which are otherwise difficult to assess [7]. In particular, via the invariant properties of datasets captured by TDA, it is possible to isolate and identify groups of points exhibiting similar features.

4.4   Network Theory for Big Data

As discussed above, data has intrinsic relationships captured by its form and type. Clearly, not all such relationships represent meaningful information, and one of the challenges is to remove redundant and inaccurate information [13]. Whilst DTA identifies the global geometrical properties of datasets, network theory offers a more direct method to the analysis of mutual relationships among data. In particular, the network topology plays a crucial role in terms of connectedness of the nodes of the network.

Network theory has increasingly drawn attention from a variety of interdisciplinary research fields, such as mathematics, computer science, biology and the social sciences. In general, networks consist of a collection of nodes called the node set $V = \{v_i\}_{i=1}^{n}$ which are connected as specified by the edge set $E = \{e_{ij}\}_{i \neq j=1}^{n}$ [8]. There are several ways to define networks from data, which can be further analysed according to their properties. Therefore, a crucial step is the identification of the topological structure associated with such network to enable a full dynamical and statistical investigation of the data set(s) modelled by it.

The investigation of network data has been considered in a wide range of real-world complex settings [9], where it has been found that the majority of network analyses ignore the network topology. In this section, the applications of network topology to mining data are discussed to facilitate the visualising structural relationships that cannot be computed using existing methods.

4.4.1   Scale-Free, Small-World and Random Networks

The network topologies that have been mostly investigated include scale-free, small-world and random networks [9]. The interest that they have been attracted has been motivated by real-world scenarios, which can be successfully modelled by these types of networks [14, 15].

Having said that, scale-free and small networks have been at the centre of current research since their structure follows specific rules, which exhibit properties utilised in the investigation of influence between their nodes [10].

Scale-free networks can be found in a variety of contexts, such as the World Wide Web links, as well as biological and social networks [8]. Furthermore, as data extraction and analysis techniques are being continuously improved, more instances of such networks will become available. This type of network is characterised by a degree distribution which follows a power law, or in other words, the fraction p_k of nodes in the network having degree k, or k connections to other nodes, can be approximated, for large values of k, as

$$p_k \approx k^{-\gamma} \tag{4.1}$$

where γ is a parameter which has been empirically shown to be usually in the range $2 < \gamma < 3$ [8]. Properties such as the preferential attachment [8] have been investigated to explain conjectured power law degree distributions in real-world networks.

One of the most important properties of scale-free networks is the relatively high probability to have *hubs*, or nodes with a degree that is much bigger than the average node degree. Such hubs are clearly relevant not only in assessing the topological properties of the corresponding networks but also in determining how the "flow of information" is affected. As a simple example, consider highly influential individuals in social networks. These tend to have many connections (Fig. 4.8).

Random graphs are described by a probability distribution, which specifies the existence of the edges between any two nodes [8]. Their applicability varies across several areas in which complex networks are investigated, and as a consequence, a large number of random graph models have been analysed, to address the different

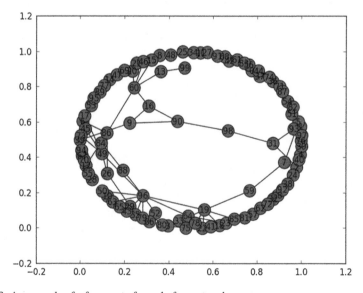

Fig. 4.8 An example of a fragment of a scale-free network

types of complex networks. An important property of random networks that will be exploited in this paper is that the fraction p_k of nodes with degree k is approximately given by

$$p_k \approx \frac{z^k e^{-z}}{k!} \tag{4.2}$$

where $z = (n - 1)p$.

Small-world networks [8] have also been extensively investigated due to their applicability to a variety of real-world models. They exhibit the property that most nodes have a relatively small degree, and they can be reached from any other node by a small number of steps. Small-world networks were first introduced following the famous small-world experiment carried out by Milgram, in which the average path length for social networks among people in the USA was analysed [11].

Specifically, a small-world network is defined to be a network where the typical distance d between two randomly chosen nodes grows proportionally to the logarithm of the number of nodes n in the network, or in other words

$$d \propto \log n \tag{4.3}$$

Historically, small-world networks were defined as a class of random graphs [8] and identifiable according to their clustering coefficient and average shortest path length. On the other hand, random graphs exhibit a small average shortest path length along with a small clustering coefficient. Watts and Strogatz [8] found out that in fact many real-world networks have a relatively small average shortest path length, but also a clustering coefficient significantly higher than expected by random chance. This has lead them to introduce a novel graph model, currently named the *Watts and Strogatz model*, with both a small average shortest path length and a large clustering coefficient.

In [10], a method to assess influence among concepts, based on specific statistical properties, is discussed. In particular, this approach exploits the approximation of the networks associated to datasets, with a topologically reduced structure. More specifically, when a dataset is shown to follow a scale-free topology, then its different components exhibit well-defined relations, such as the preferential attachment property [8]. Furthermore, the dynamical properties of topologically reduced networks lead to the understanding of the level of influence between any two nodes.

The ability to determine probabilistic information captured by networks associated with datasets is of crucial importance in the design of any knowledge extraction system. However, there are a variety of complex challenges to overcome, especially when large and unstructured data sets are to be addressed. In [12], the authors introduce a method to identify and assess the joint probability of concepts embedded onto networks defined from unstructured large data sources. The preliminary results show great potential, suggesting that this approach can be subsequently integrated into a system to extract, identify and assess intelligence and ultimately facilitating the decision-making process [12].

4.5 Conclusions

In this chapter, an overview of the most promising emerging approaches to Big Data is discussed. The combination of their strong theoretical foundations and computational efficiency has shown to provide useful tools in knowledge representation and discovery. Furthermore, this suggests that Big Data research requires an integrated effort from a variety of multidisciplinary research fields to tackle its increasing complexity.

References

1. Mayer-Schönberger V, Cukier K (2013) Big data: a revolution that will transform how we live, work, and think. Houghton Mifflin Harcourt, Boston
2. James IM (1999) Preface. In: James IM (ed) History of topology, North-Holland, Amsterdam, p v. doi:10.1016/B978-044482375-5/50000-6. ISBN 9780444823755
3. Jänich K (1984) Topology. Springer, New York
4. Lum PY, Singh G, Lehman A, Ishkanov T, Vejdemo-Johansson M, Alagappan M, Carlsson J, Carlsson G (2013) Extracting insights from the shape of complex data using topology. Sci Rep 3, 1236
5. Dey TK, Fan F, Wang Y (2013) Graph induced complex on point data. In: Proceedings of the 9th annual symposium on computational geometry. Rio de Janeiro, Brazil, June 17–20
6. Warner FW (1983) Foundations of differentiable manifolds and Lie groups. Springer, New York
7. Carlsson G (2009) Topology and data. Bull Am Math Soc 46(2):255–308
8. Albert R, Barabási AL (2002) Statistical mechanics of complex networks. Rev Mod Phys 74:47
9. Trovati M (2015) Reduced topologically real-world networks: a big-data approach. Int J Distrib Syst Technol (IJDST), 6(2):13–27
10. Trovati M, Bessis N (2015) An influence assessment method based on co-occurrence for topologically reduced big data sets. Soft Comput. doi:10.1007/s00500-015-1621-9
11. Milgram S (1984) The individual in a social world. McGraw-Hill, New York
12. Trovati M, Bessis N, Palmieri F, Hill R Extracting probabilistic information from unstructured large scale datasets. IEEE Syst J (under review)
13. Duda R, Hart PE (1973) Pattern classification and science analysis. Wiley, New York
14. Trovati M, Bessis N, Huber A, Zelenkauskaite A, Asimakopoulou E (2014) Extraction, identification and ranking of network structures from data sets. In: Proceedings of CISIS. Birmingham, UK, pp 331–337
15. Trovati M, Asimakopoulou E, Bessis N (2014) An analytical tool to map big data to networks with reduced topologies. In: Proceedings of InCoS. Salerno, Italy, pp 411–414

Part II
Applications

Chapter 5
Integrating Twitter Traffic Information with Kalman Filter Models for Public Transportation Vehicle Arrival Time Prediction

Ahmad Faisal Abidin, Mario Kolberg, and Amir Hussain

Abstract Accurate bus arrival time prediction is key for improving the attractiveness of public transport, as it helps users better manage their travel schedule. This paper proposes a model of bus arrival time prediction, which aims to improve arrival time accuracy. This model is intended to function as a preprocessing stage to handle real-world input data in advance of further processing by a Kalman filtering model; as such, the model is able to overcome the data processing limitations in existing models and can improve accuracy of output information. The arrival time is predicted using a Kalman filter (KF) model, by using information acquired from social network communication, especially Twitter. The KF model predicts the arrival time by filtering the noise or disturbance during the journey. Twitter offers an API to retrieve live, real-time road traffic information and offers semantic analysis of the retrieved twitter data. Data in Twitter, which have been processed, can be considered as a new input for route calculations and updates. This data will be fed into KF models for further processing to produce a new arrival time estimation.

5.1 Introduction

A number of studies have shown that the majority of arrival time prediction models are based on historical arrival patterns and/or other explanatory variables correlated with the arrival time. The explanatory variables used in previous studies include historical arrival time (or travel time), schedule adherence, weather conditions, time-of-day, day-of-week, dwell time, number of stops, distance between stops, and road-network condition [1, 2]. The collection and transmission of such variables has largely been made possible using emerging technologies, including wireless communication, automatic vehicle location (e.g., GPS), and other sensing technologies. The effect of congestion has been treated in a manner that varies between models. For example, some have used traffic properties like volume and

A.F. Abidin (✉) • M. Kolberg • A. Hussain
Computing Science and Mathematics, University of Stirling, Stirling, Scotland, UK
e-mail: faa@cs.stir.ac.uk; mko@cs.stir.ac.uk; ahu@cs.stir.ac.uk

© Springer International Publishing Switzerland 2015
M. Trovati et al. (eds.), *Big-Data Analytics and Cloud Computing*,
DOI 10.1007/978-3-319-25313-8_5

speed from simulation results, while some have clustered data into different time periods. Historical data-based models were used in areas where traffic congestion is less, because the models assumed cyclical traffic patterns [ref]. Kalman filtering techniques and artificial neural network (ANN) approaches were used mostly in urban areas [3]. Kalman filtering models can be applied on-line while the bus trip is in progress due to the simplicity of calculation.

There are many algorithms grounded in mathematical theory and/or statistical models that have been proposed for bus travel time prediction. However, there is a gap evident in these algorithms. One particular issue is the question of how the algorithm will receive and incorporate live, real-time traffic event information. Another issue is how this information can be passed and be made available, so that prediction approaches can make use of it. This paper proposes that social network communication can be used to make necessary information available to arrival time prediction algorithms. Social network communication is a novel way to collect and include current road condition information that can be used as additional approach (existing approaches, e.g. using GPS or other (road) sensors to detect the numbers of cars on the road and speed of travel) to improve accuracy of vehicle arrival time prediction. In addition, this approach allows for the identification of unexpected traffic events and the inclusion of this new, real-time information as part of potential route calculations and updates. This provides up-to-the-minute information during journeys, which may not have been available when travel was initially planned or started. In this situation, social networks can play a pivotal role as an input to a scheduling model.

The Kalman filter, also known as linear quadratic estimation, is an algorithm that uses a series of measurements observed over time, containing noise, and produces estimates of unknown variables. The pressing issue in Kalman filtering-based models is the type of parameters used as inputs. For example, in the scenario of an accident occurring, how can the delayed information relating to this incident be input into a Kalman filtering model for further process? In real-world situations, the dynamic nature of information needs to be taken into account to improve accuracy and feasibility. In this situation, social networks, particularly Twitter, can improve the accuracy of information by providing real-time input on current incidents. The nature of Twitter, which provides a platform for people to communicate about work, weather, travel, or any topic of interest, can be applied to Kalman models in terms of feeding dynamic information. Twitter's networking component allows users to make connections with friends of friends, and this dynamic can lead to serendipitous professional or personal relationships with other Tweeters. Twitter creates a new channel of communication and also facilitates a new way of interacting people. This also can be leveraged in Kalman models to improve accuracy of information. Contrary to major commercial offerings such as Tomtom, which relies on data from its large user base, the system proposed here does not require a very large number of simultaneously active users. Instead, it uses information from social networks which is independent of the traffic application.

5.2   Communication Platform on Twitter

Messages in Twitter ("tweets") are straightforward. Twitter is a microblogging site where users exchange short, maximum 140-character messages. There are a number of special characters used including @, RT, and #. Messages starting with "@" are called @replies because they reply directly to someone else's tweet. Taking user name "aragorn" as an example, if a message begins with "@aragorn", it meant that the owner of the "aragorn" account will see it, even if "aragorn" is not following the message creator's account. The @username can also appear in a message, allowing that user to see the message. The @ function is the most widely used Twitter symbol. Twitter also supports a mechanism to connect one message to another. This connection between specific messages is often called "Threading" and helps people see the context of the conversation.

The retweet function is used when a message begins with the Twitter symbol "RT". This indicates that a person is reposting someone else's message. RT is a clue that the message may be worth additional attention, because somebody thought it worthy to pass on. Messages can also contain arbitrary labels beginning with the "#" Twitter symbol, known as a hashtag. It is a way of categorising messages, allowing one to search messages containing a particular tag. Twitter messages are public (if broadcast on a Twitter wall), so the search will cover all messages. This creates a collective snapshot of what people are saying about a given topic.

5.3   Communication for Data Collection on Twitter

Twitter is a platform that can handle a wide range of communication such as the discussion on certain issues between individual and groups. The communication strategies in Twitter are to study and investigate the communicative patterns within hashtags, keywords and data sets which are extracted from daily interactions. Upon providing a hypothetical description of the Twitter communication analysis [4], customising the approach and methodology to carry-out particular analyses is necessary. In addition to the general findings about Twitter's communication structure, a large amount of data can be used in order to acquire a better understanding of certain issues or events [5] and to predict specific real-time events, such as massive traffic congestion on a particular road in Edinburgh.

In essence, Twitter communication analysis can be performed by employing metadata provided in application programming interface (API). The use of metrics for Twitter communication analysis is uncomplicated and builds upon the communication data collected through the Twitter API. The Twitter API [6] provides a streaming API, which provides real-time access to Tweets in sampled and filtered forms, making it possible for Twitter's API to extract live real-time interaction data from Twitter users. Real-time access or real-time streaming allows the Twitter API to access any type of tweets or retweets broadcast by twitter users. The fascinating aspect of real-time streaming is that it is able to perform minute by minute live updates of the data set. This allows for potential improvement of the decision-making process by using the most up-to-date data for further processing. In the context of this paper, receiving real-time data is intended to improve bus arrival time.

The streaming API is grounded on a push-based approach [6]. This push-based approach allows data to constantly stream upon request. Users can manipulate the Twitter API to receive live data on a constant basis and at the same time process data for a specific use. This streaming data is provided as a live feed; as such, particular data can be extracted as soon as it is tweeted. Studies of real-time events, such as road traffic incidents, require researchers to rigorously establish a stream for collection and sorting of data, in order to compile an analytically useful set.

In this research, we make use of Tweetbinder (Twitter API Tool), which has the ability to generate deep analytics of Twitter users based on various filters, such as keywords, hashtags, pictures, textual tweets, and retweets. The Binder is a prime component in Tweetbinder that was developed to tackle two problems: textual organisation and statistics. A textual organiser makes it possible to sort hashtagged tweets into categories (binders), thus making it easier to organise tweets from a particular event (or hashtag) depending on set criteria. In each binder, tweets are separated into textual tweets, pictures, links, and retweets. This allows researchers to conduct granular level data analysis. The concepts introduced in this paper provide a pragmatic set of analytical tools with which to study information retrieved from Twitter communications. The specific metrics that relate to particular Twitter communicative contexts may also be leveraged. A practical method available for research use (the Twitter API) is employed to extract requested information.

5.4 Event Detection and Analysis: Tweets Relating to Road Incidents

Information retrieval is a popular technique provided in the Twitter API (Tweetbinder (Fig. 5.1)). It has the ability to search, retrieve and analyse based on hashtags, keywords and terms [7]. Information retrieval is very appropriate for specific events such as real-time incident detection and analysis of Twitter messages for the purpose of critical and noncritical situations. Twitter interactions have also previously been employed as a medium for notification of road transportation incidents [8]. Eric Mai et al. [8] evaluate the use of data from public social interactions on Twitter as a potential complement to road traffic incident data. The data correlated to road traffic incidents can be compared to incident records from authorities based on the matching time period.

Twitter provides a platform for notifying users of real-time events, such as traffic incidents. There are a number of nongovernment bodies that focus on notification and information dissemination relating to road traffic incidents, whether

Fig. 5.1 Twitter API (Tweetbinder) has the capability to search and retrieve information which is requested

positive or negative, such as Edinburgh Travel News (using hashtag #edintravel) and Traffic Scotland. Comparing tweets relating to road traffic incidents can improve consistency and avoid scam tweets. Tweets represent a witness perspective and a test case for using public broadcast user-generated content to send notifications about road conditions.

Using Twitter as a data source for traffic incidents has geographically context-sensitive versatility: Twitter content can be analysed in a repeatable way to provide similar information anywhere with sufficiently dense Twitter usage. Twitter information is user generated and so can be considered a source of user centric data, as road users' experiences can vividly describe traffic incidents. The authorities have installed road cameras as a way to identify traffic issues; however, problems arise as cameras cannot be installed on all roads due to high costs. Thus, road users can use Twitter to disseminate information about incidents and can include live real-time information as part of route calculations and updates. This provides information during journeys that may not have been available when travel was initially planned or started. In this situation, social networks can play a pivotal role as an input to a scheduling model.

5.4.1 Twitter Data: Incident Data Set

Twitter data were collected from Twitter's third-party real-time streaming application, Tweetbinder, through an information retrieval technique, as previously discussed. Various filters can be set on these data streams to capture tweets within a particular geographic area, or containing certain terms. Data for this study was filtered to contain keywords that relate to traffic incidents. These keywords are accident, crash, traffic, road, carriageway, delays, congestion, lane, car, and cars. An extremely large quantity of real-time data is available at any given time on Twitter. However, the keyword filter is applied to filter the global Twitter stream. Another problem is that Twitter data can potentially contain global data of all matching tweets. In order to mitigate this issue, the study employed location (a focus on Edinburgh city), temporal, and time series metrics. Employing these metrics can dramatically reduce the size of the data set. In addition, Tweetbinder has the capability to reduce the size of the data through the binder function. This can sort information based on particular roads, incidents, and timestamps and is very useful to detect particular events that are requested.

Twitter has a location-based component, which has the capability to broadcast a tweet's location. Location-based component can be leveraged for twitter users to employ the concept of 'at the location'. 'At-the-location' concept allows to disclose the location where the content is created. With this, Twitter users are able to broadcast, in real-time, any event they witness. In this regard, one issue that Twitter has problems with currently is the lack of messages that contains the incident location. This problem indirectly affects the ability to probe messages with respect to traffic incidents. However, this issue does not impact the #edintravel hashtag. Each message broadcast in #edintravel contains the specific location of incidents. In addition, #edintravel messages clearly state where the incident happened. Hence, #edintravel can be employed as a central information source, which contributes to dissemination of information concerning road traffic incidents throughout Edinburgh.

Incident data were retrieved via Twitter's API, from the #edintravel live incident feed. This feed is publicly accessible and displays all active traffic incidents in Edinburgh at any given time. Only live incidents and those involving a particular day are available. These incident locations are labelled with a road name, and sometimes together with the time, the incidents were cleared (Fig. 5.2). In addition, a number of tweets also notify effects to, and from, other routes or roads that will be affected as a consequence of the incident (Fig. 5.3). Interestingly, most of the data from #edintravel are acquired from crowdsourcing interactions on Twitter. In addition, transportation service companies feed information using this hashtag. As shown in Fig. 5.4, bus companies feed information about bus diversions via Queen Margaret University and the A1 in Edinburgh, as they cannot pass through Newcraighall road due to road closures. This kind of information gives vital early notifications to those commuting by bus in that area, which cannot be provided in grid positioning system (GPS) devices. In addition, #edintravel can be expressed as data centric tweets, which can be leveraged to acquire information relating to road incidents throughout Edinburgh. In Fig. 5.1, information on incidents throughout Edinburgh is retrieved from #edintravel; the incidents took place throughout Edinburgh between 23 April 2015 and 25 April 2015.

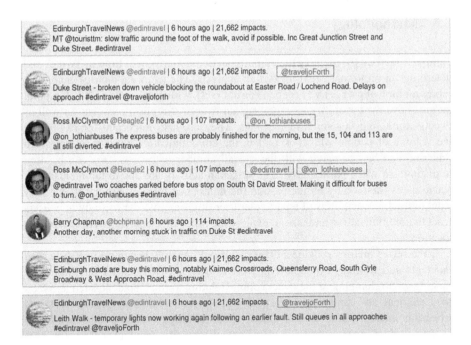

Fig. 5.2 Information on incidents in Edinburgh city

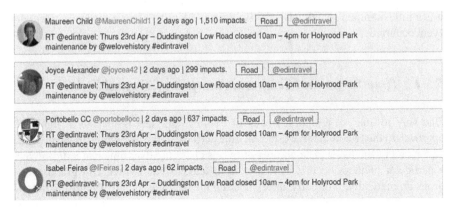

Fig. 5.3 Early notifications pertaining to road closure in Duddingston Low Road, Edinburgh

Fig. 5.4 Information on bus diversions due to road closures

5.5 Methodology

Temporal metrics [9] identify the communication patterns over time in Twitter and determine an important event for further analysis. The temporal metrics enable the representation of user activity (depend on data set), with graph spikes [10]. Temporal metrics have the capability to track original tweets, @replies, retweets, as well as tweets containing keywords or hashtags. They are also important for examining the volume of people who took part in the communication activity at any given time and can be compared with the overall number of tweets. This can be used to distinguish when particularly large spikes in activity are due to the inclusion of active users (marked by an increase in the number of tweets). Similarly, one may want to track the activities of a particular user or group in order to study how users react to communication and also to explore the types of users that tweet in multiple locations.

Time series metrics provide a clear picture of people's reaction to specific road incidents and to change people's interest over time. This can be interpreted by looking at the changing of the tweet volume relating to road incident, such as a fatal road accident, and can reveal when people are interested about it (post a tweet) and when this interest starts to end (nobody tweet relating to specific incident). Peaks of the tweet volume can be used to indicate the specific incident within a broader event that generated the most interest. Twitter is important in order to perceive and understand crowdsourcing opinion on certain issues. As it is an open application that allows all data to be made public if the tweet is posted on a user's wall and as it is a time-stamped social network, Twitter can identify when a certain incident or event occurred.

5.5.1 Time Series and Temporal Analysis of Textual Twitter

One way to analyse temporal trends of Twitter data sets is to take a graphical approach to the tweets at three different time periods: at the beginning, in the middle, and at end of the tweet. The number of tweets changing over time can indicate that a particular incident occurs, as it is hotly discussed via Twitter. If a graph spike drops dramatically, it directly shows that the incident has ended. Therefore, in this section we use the graphical approach and described analysis to illustrate an event's description.

The time series approach is very important in tracking an event that is discussed on twitter. In order to conduct an analysis using this approach, the context or trend of discussions must correlate with the event. The time series approach will be able to distinguish other events that may occur elsewhere. This approach allows researchers to study the shape of the graph at a specific period of time. By using a time series approach, each tweet can be tracked based on the requested time. This will indicate spikes of various sizes on the graph. Although the largest spikes garner the most attention, this study considers each spike important. The bus may pass through an

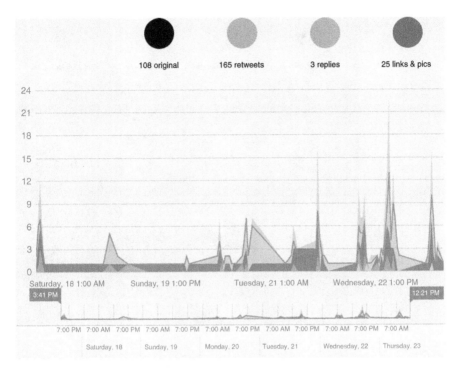

Fig. 5.5 The graph represents tweets in A720, Edinburgh

area rarely travelled by road users in Edinburgh; thus, people may want to know the situation and the condition of the road so they can estimate the time the bus will arrive at its destination. The time interval used is in minutes; therefore each point on the graph represents the number of topic-relevant tweets gathered within minutes in the Twitter API, Tweetbinder (discussed in previous sections). The time frame on Tweetbinder only focuses on spikes; therefore, the time interval in Fig. 5.5 is not consistent. Graph in Fig. 5.5 is generated from a data set which is filtered by employing binder component in Tweetbinder. The tweets about the conditions on a carriageway (A720 – E/B at Sheriffhall roundabout) can be searched from April 18th to April 23rd, 2015. The size of a spike is either large or small, depending on the amount and time period of tweets discussing an incident. The large spike illustrates something interesting, leading many people to tweet. Large spikes also reflect an incident occurring in a long period of time. The small spikes show a road incident in Edinburgh, which may occur in a place rarely travelled by the public. As such, only a few people tweet about it. In the following analysis, the small spikes also describe an incident that occurred in a very short period of time. Hence, only a few people saw the incident occurred and could tweet about it.

This time frame determines if adverse road conditions influence the graph's spike. A large spike on the graph (Fig. 5.6) illustrates something interesting that causes people to tweet, which is the road congestion on road A720, Sheriffhall

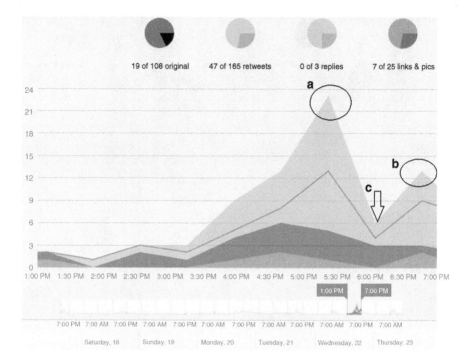

19 of 108 original   47 of 165 retweets   0 of 3 replies   7 of 25 links & pics

Fig. 5.6 Large spikes can provide specific evidence related to the incident time

roundabout. Large spikes reflect incidents which occurred in a quite long period of time. At its peak, more than 19 original, relevant tweets and 47 retweets were sent in less than 3 h. The peak in interest happened significantly after incidents began on road A720, Sheriffhall roundabout. The graph can help in identifying the overall trends of interest, as well as particular points of interest. If the graph focuses on time-limited topic, as shown here, then one observes an increase in the number of tweeting at early time points, when the topic began to attract interest. Content analysis of tweets in the early stages may also show people's initial reaction. At this stage, it is important to be aware and track first tweets. Sometimes, people share the same tweet, but disseminate it at different times. Thus, to overcome this problem, we employed Tweetbinder to organise the data sets into segments by leveraging 'binder' in Tweetbinder. By creating 'binders' based on customised filters, a user can track in real time how Twitter followers react to incidents and events. Subsequently, a 'binder' can able to identify the original tweet for a particular incident through an arrangement of tweets in that binder. The initial increase in tweeting volume points to the time (3.15 p.m.) at which the incident's discussion gains interest. Tweet's content at the initial stages indicates initial reactions to the tweet. Discussions become increasingly culminate (at approximately 5.15 p.m.) as 23 people tweeted about the event. The peak of the discussion can be known by looking at the peak of

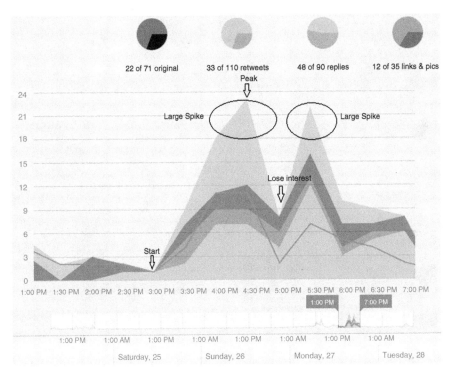

Fig. 5.7 Large spike indicating a specific event of interest

the graph spikes (Fig. 5.6a). This can be interpreted as the topic of discussion getting more attention in the public consciousness. The graph gradually decreases in volume of tweets (at approximately 6.12 p.m., only six tweets (Fig. 5.6c)) highlighting when people started losing interest in that discussion (suggesting the incident has ended). A second spike at point (b) could be interpreted as there being new information about the event, which gets attention. Alternatively, it could be about a new event. In order to tackle this problem and to avoid confusion, the tweet content may need to be checked.

If the topic is continued for a long time, such as interest in severe road congestions causing major travel disruption, then a comparison of tweet content analysis between the start and end of the period is investigated. This may reflect to change in interest as they take place. In Figs. 5.7 and 5.8, spikes indicate specific event of interest for the topic. In order to identify the interest of specific event, the tweets that are retrieved from the spike must be referred. This can be achieved through content analysis, if necessary.

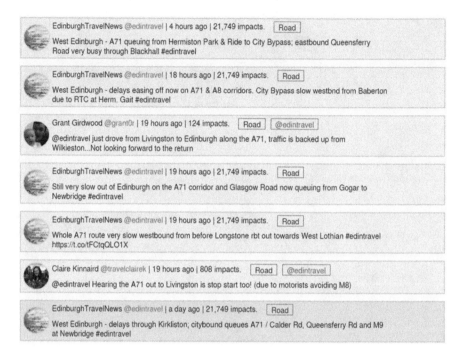

Fig. 5.8 Information relating to large spikes in Fig. 5.7

5.6 Proposed Refined Kalman Filter (KF) Model-Based System

Kalman filters are an estimation approach. That is, they infer values of parameters from observations which may be noisy, inaccurate, and uncertain. Importantly, unlike many other approaches, Kalman filters are recursive and hence can take new observations into account as they arrive. With this, Kalman filters can be executed at *runtime* of the system under observation. The algorithm is referred to as a 'filter' as calculating the estimate from noisy input data essentially 'filters out' the noise.

Kalman filters estimate a process by estimating the process state at a given time and then obtaining feedback in the form of noisy measurements. Generally, the equations for the Kalman filter fall into two groups: time update equations and measurement update equations. Time update equations are responsible for predicting forward (in time), using the current state and error covariance estimates to obtain the a priori estimates for the next time step. The measurement update equations are responsible for the feedback, i.e. for incorporating a new measurement into the a priori estimate to obtain an improved a posteriori estimate. The time update equations can also be thought of as predictor equations, while the measurement update equations can be thought of as corrector equations. Indeed, the final estimation algorithm resembles that of a predictor-corrector algorithm for solving numerical problems.

A Kalman model implies the state of a system at a time $k + 1$ developed from the previous state at time k. This can be expressed by the following state equation:

$$x_{k+1} = \alpha x_k + \beta u_k + w_k \tag{5.1}$$

Here, x_{k+1} is the state vector containing the terms of interest at time $_{k+1}$. u_k is the vector with the control inputs at time $_k$. α is the state transition matrix which applies the effect of system state parameters at time $_k$ to the state at time $_{k+1}$. β is the control input matrix which applies the effects of the control input parameters (u_k) to the state vector. w_k is a vector containing process noise terms for each value in the state vector.

Measurements of the system are performed according to the formula

$$y_k = \mu x_k + z_k \tag{5.2}$$

where y_k is the vector of measurements, μ is the transformation matrix which maps the state vector parameters to the measurements, and z_k is the vector which contains the measurement noise for each element in the measurement vector. This measurement formula is also referred to as output equation.

Consequently, Kalman filter (KF) model can be used to estimate the position of a vehicle by inputting the vehicle speed into the KF algorithm. The addition of state constraints to a KF model can significantly improve the filter's estimation accuracy [11]. In this sense, the addition of information that is input linearly into the model may produce significant benefits. KF models can theoretically deliver the best and most up-to-date results when they have continuous access to dynamic information [12]. Many existing models therefore make use of dynamic information. However, the performance of these models can often suffer due to issues with accounting for scenarios that involve rapidly updating, real-world information. In these cases, constraints may be time varying or non-linear. Many road users rely on navigation systems to navigate and estimate the duration of their journeys. However, it is not always possible to handle some events solely by navigation systems' information. For instance, if there is an accident, and it is known that it will take 30 min or more to clear, a navigation system cannot detect or incorporate this information because it provides and predicts an arrival time based on GPS satellites. As discussed, social networks have the ability to provide plausible real-time information regarding road traffic, which could potentially improve the accuracy of arrival time prediction.

Intelligently/automatically selecting the best source of external traffic condition information from social networks for input into the KF model can produce improved traffic prediction results. This is achieved by comparing conventional GPS-based Traffic Management Systems (TMS) with new social media information sources. As this paper previously noted, the external/delay information can initially be 'linearly' added to determine total KF, based on arrival estimation times. For instance, if a KF model estimates the arrival time without external delay information to be 80 min, and there is delay information from social media of 20 min, then the estimated arrival time will be $= 80 + 20 = 100$ min.

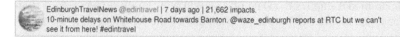

Fig. 5.9 Linear information notified 10-min delay on Whitehouse Road

While travel time data can be obtained from various sources – such as loop detectors, microwave detectors, and radar – it is unrealistic to assume that the entire road network is covered by such data collection devices. Referring to Fig. 5.9, information from EdinburghTravelNews such as '10-min delays on Whitehouse Road towards Barnton' can be leveraged. This can be input linearly into KF models. Other social media data (see: Fig. 5.6) can also be fed into KF models. Previously constructed KF models [12, 13] focused on speed and position. Hence, a 30-min delay will impact vehicle speed and position. Due to this, the speed, calculated in Eq. 5.3, will be set to 0 and a 30-min delay set in Eq. 5.6. The key component of the previously configured KF model [19] is outlined below:

$$x = \begin{pmatrix} \text{Position} \\ \text{Speed} \end{pmatrix} \tag{5.3}$$

The system model is then set as follows:

$$x_{k+1} = Ax_k + w_k \tag{5.4}$$

$$y_k = Hx_k + v_k \tag{5.5}$$

$$A = \begin{pmatrix} 1 & \Delta t \\ 0 & 1 \end{pmatrix}, H = \begin{pmatrix} 0 | 1 \end{pmatrix} \tag{5.6}$$

Similarly to the general Kalman filter equations above, x_k represents the state variable, y_k the measurement variable, A the state transition matrix, H the state measurement matrix, and v_k the measurement noise, defined as any unusual disturbance during a journey. Further, A describes how the system changes over time. Δt is the interval of the time measuring position. Equation 5.4 predicts the state at the next time step for system model A. Subscript k indicates that the KF model is executed in a recursive manner. The state measurement matrix H is used in Eq. 5.6 to predict the position based on measured velocity. Designing the optimum system model is an art and difficult to calculate precisely. The system model has to rely on the experience and capability of the Kalman filter. The additional w_k is introduced in Eq. 5.4 to show the noise taken into account and will give effects of the state variable.

To demonstrate the accuracy of prediction, and that the simulated system results correspond to real-world conditions, a simulation of urban mobility (SUMO) is used [14, 15] to validate the KF models. SUMO is a road traffic simulation which is capable of simulating real-world road traffic using digital maps and realistic traffic models. This is fully discussed in [12] and indicates that appropriately constructed KF models correspond to real-world data. In addition, the estimation arrival time in KF models is more accurate if more information is fed into the model.

5.7 Conclusion

KF models are well-established tools with which to estimate vehicle arrival time. The strength of KF models is their ability to predict or estimate the state of a dynamic system from a series of noisy measurements. Based on this strength, additional credible information from social networks can be leveraged to feed into KF models. The nature of social networks, especially Twitter, provides real-time information and is vital to get up-to-date news related to road traffic. Twitter API (Tweetbinder) is an application that can be used effectively to retrieve and analyse relevant requested information. The ability of Twitter API to sort information based on temporal and time series metrics can be leveraged to reduce the data size and focus on particular data. Data that has been sorted will feed into KF models to produce a new and more accurate estimation time.

For future work, Twitter information can be semantically mined using sentic computing based sentiment and opinion mining techniques [16, 17], to infer the context and true meanings from users' messages.

References

1. Tongyu Z, Dong J, Huang J, Pang S, Du B (2012) The bus arrival time service based on dynamic traffic information. In 2012 6th international conference on application of information and communication technologies (AICT), Tbilisi. IEEE, pp 1–6
2. Williams BM, Hoel LA (2003) Modeling and forecasting vehicular traffic flow as a seasonal ARIMA process: theoretical basis and empirical results. J Transp Eng 129(6):664–672
3. Lesniak A, Danek T (2009) Application of Kalman filter to noise reduction in multichannel data. Schedae Informaticae 17:18
4. Bruns A, Stieglitz S (2013) Metrics for understanding communication on Twitter. Twitter Soc 89:69–82
5. Tumasjan A, Sprenger TO, Sandner PG, Welpe IM (2010) Predicting elections with Twitter: what 140 characters reveal about political sentiment. ICWSM 10:178–185
6. Gaffney D, Puschmann C (2014) Data collection on Twitter. In: Twitter and society. Peter Lang, New York
7. Tao K et al (2014) Information retrieval for Twitter data. In: Twitter and society. Peter Lang, New York, pp 195–206
8. Mai E, Rob H (2013) Twitter interactions as a data source for transportation incidents. In Proceedings of the transportation research board 92nd annual meeting, Washington, DC, no. 13–1636
9. Steur RJ (2015) Twitter as a spatio-temporal source for incident management. Thesis, Utrecht University. Utrecht University Repository, Utrecht (Print)

10. Kumar S, Morstatter F, Liu H (2014) Visualizing Twitter data. In: Twitter data analytics. Springer, New York, pp 49–69
11. Simon D (2010) Kalman filtering with state constraints: a survey of linear and nonlinear algorithms. IET Control Theory Appl 4(8):1303–1318
12. Abidin AF, Kolberg M (2015) Towards improved vehicle arrival time prediction in public transportation: integrating SUMO and Kalman filter models. In: IEEE 17th international conference on modelling and simulation UKSim-AMSS, University of Cambridge
13. Abidin AF, Kolberg M, Hussain A (2014) Improved traffic prediction accuracy in public transport using trusted information in social networks. In: 7th York doctoral symposium on computer science & electronics. University of York
14. Behrisch M, Bieker L, Erdmann J, Krajzewicz D (2011) SUMO–Simulation of Urban Mobility. In: The third international conference on advances in system simulation (SIMUL 2011), Barcelona
15. Krajzewicz D, Erdmann J, Behrisch M, Bieker L (2012) Recent development and applications of SUMO–simulation of urban mobility. Int J Adv Syst Meas 5(3/4):128–138
16. Cambria E, Hussain A (2012) Sentic computing: techniques, tools and applications. In: SpringerBriefs in cognitive computation. Springer, Dordrecht, 153 p
17. Poria S, Cambria E, Howard N, Huang GB, Hussain A (2015) Fusing audio, visual and textual clues for sentiment analysis from multimodal content. In: Neurocomputing. Elsevier Academic Press, Netherlands, pp 1–9

Chapter 6
Data Science and Big Data Analytics at Career Builder

Faizan Javed and Ferosh Jacob

Abstract In the online job recruitment domain, matching job seekers with relevant jobs is critical for closing the skills gap. When dealing with millions of resumes and job postings, such matching analytics involve several Big Data challenges. At CareerBuilder, we tackle these challenges by (i) classifying large datasets of job ads and job seeker resumes to occupation categories and (ii) providing a scalable framework that facilitates executing web services for Big Data applications.

In this chapter, we discuss two systems currently in production at CareerBuilder that facilitate our goal of closing the skills gap. These systems also power several downstream applications and labor market analytics products. We first discuss Carotene, a large-scale, machine learning-based semi-supervised job title classification system. Carotene has a coarse and fine-grained cascade architecture and a clustering based job title taxonomy discovery component that facilitates discovering more fine-grained job titles than the ones in the industry standard occupation taxonomy. We then describe CARBi, a system for developing and deploying Big Data applications for understanding and improving job-resume dynamics. CARBi consists of two components: (i) WebScalding, a library that provides quick access to commonly used datasets, database tables, data formats, web services, and helper functions to access and transform data, and (ii) ScriptDB, a standalone application that helps developers execute and manage Big Data projects. The system is built in such a way that every job developed using CARBi can be executed in local and cluster modes.

6.1 Carotene: A Job Title Classification System

In many diverse domains such as bioinformatics and e-commerce (among others), content is organized according to concept hierarchies and taxonomies. A few examples of well-known taxonomies are Amazon.com, Ebay.com, the United Nations Standard Products and Services Code (UNSPSC), and Wikipedia (among

---

F. Javed (✉) • F. Jacob
Data Science R&D, CareerBuilder, Norcross, GA, USA
e-mail: faizan.javed@careerbuilder.com

© Springer International Publishing Switzerland 2015
M. Trovati et al. (eds.), *Big-Data Analytics and Cloud Computing*,
DOI 10.1007/978-3-319-25313-8_6

others). For large e-commerce websites, accurate item classification provides more relevant recommendations and a better item catalog search and browsing experience for end users. In many highly competitive e-commerce markets, an improved end-user experience can potentially result in an increase in sales transactions and repeat customers. Most of these websites handle Big Data and have deployed large-scale systems that can classify millions of items to thousands of categories in a fast, efficient, and scalable manner.

In the online recruitment domain, CareerBuilder (CB) operates the largest online job site in the USA with more than 24 million unique visitors a month (as of March 2015). One of CB's goals is to help close the skills gap. The skills gap is defined as the perceived mismatch between the needs of employers for talent and the skills possessed by the available workforce. At CB, accurate classification of job ads (job title, description, and requirements) and resumes to occupation categories is important for many downstream applications such as job recommendations and labor market analytics products. These applications contribute to CB's goal of helping job seekers find the jobs and training they need to be more competitive on the job market. The applications also deliver valuable insights to employers to help shape their recruitment strategies. Figure 6.1 shows some examples of CB applications and products that utilize Carotene.

Carotene is a large-scale job title classification system that leverages text document classification and machine learning algorithms. An automatic document classification system that uses machine learning algorithms requires documents labeled with predefined classes to create a set of training data. The training data is then used to induce a model that can generalize beyond the examples in the training data. The model is then used to classify new documents to one or more

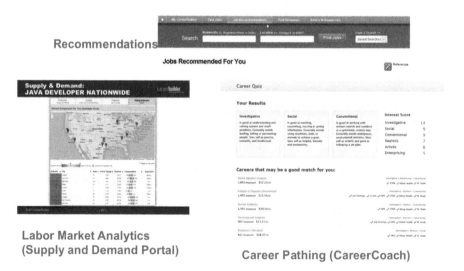

Fig. 6.1 Carotene downstream applications and products

of the predefined classes. There are many related works in both industry and academia that discuss classification systems that categorize entities to domain-specific taxonomies. In [1], a Support Vector Machine (SVM) [2] was used as both a coarse and fine-grained classifier as a cascade to classify large heterogeneous web content. A relevance vector machine (RVM)-k-nearest neighbor (kNN) classifier cascade to detect cancer from 3D medical images is discussed in [3]. In this system, data samples are first grouped into similar clusters using clustering. An RVM is used as a coarse-level classifier to prune data samples far from the classification boundary, while kNN is used as the fine-level classifier for the final classification. EBay's work on large-scale item categorization [4] uses a cascade of kNN-SVM classifiers to classify products to a homegrown taxonomy. In the online recruitment domain, LinkedIn's approach to job title classification [5] is based on a heavily manual phrase-based system that relies on the near-sufficiency property of short text. The system has offline components that build a controlled vocabulary and a list of standardized job titles as well as a component for crowdsourced labeling of training samples.

6.1.1 Occupation Taxonomies

For the online recruitment industry in the USA, the Occupational Information Network (O*NET) taxonomy is the primary source of occupation categories. O*NET is an extension of the Standard Occupational Classification (SOC) system developed by the US Bureau of Labor Statistics and was developed through the sponsorship of the US Department of Labor/Employment and Training Administration (USDOL/ETA). The O*NET taxonomy is a four-level hierarchy composed of 23 major groups, 97 minor groups, 46 broad occupations, and 1,110 detailed occupations. Table 6.1 shows the taxonomy breakdown for O*NET code 15-1132.00 which represents the *Software Developers, Applications* occupation.

The O*NET taxonomy is not granular enough for our suite of products and applications. For example, consider the niche and emerging software engineering titles such as *Big Data Engineer* and *Machine Learning Engineer*. These job titles are represented alongside more general software engineering titles such as *Java Engineer* by the *Software Developers* subcategories 15.1132 (*Applications*) and 15-1133 (*Systems Software*). The lack of fine-grained job title granularity inherent in the O*NET

Table 6.1 O*NET taxonomy groups for software developers, applications

Taxonomy group	Code and description
Major	15 – Computer and mathematical occupations
Broad	1130 – Software developers and programmers
Minor	1132 – Software developers, applications
O*NET extension	00 – Software developers, applications

and SOC taxonomies adversely affects our ability to drill down on metrics and analytics insights to the level of detail expected by our end users. These taxonomies cannot accommodate emerging job categories and titles because of long lead times between taxonomy updates. These taxonomies will be updated every 10 years.

6.1.2 The Architecture of Carotene

Classifiers for large-scale taxonomies are usually hierarchical cascade classifiers or flat classifiers. A flat classifier has to make a single classification decision that is more difficult than for a cascade classifier. A flat classifier also encompasses all leaf nodes of the taxonomy as classes that for large-scale taxonomies can be in the thousands. A hierarchical classifier is preferred for highly unbalanced large-scale taxonomies [6] because even though a hierarchical classifier has to make multiple decisions, there is a less severe imbalance at each level of the hierarchy. However, hierarchical classifiers are susceptible to error propagation because the final classification decision relies on the accuracy of previous decisions made in the hierarchy. Pruning the taxonomy can mitigate error propagation in hierarchical classifiers.

Carotene is a semi-supervised job title classification system that also has a clustering component for taxonomy discovery. The classification component is a multi-class, multi-label cascade classifier that for a query text returns a list of job titles ranked by confidence scores. Carotene only takes into consideration the top (SOC major) and O*NET code levels of the four-level O*NET taxonomy. To create the training dataset for Carotene, we collected 3.6 m job ads posted on CareerBuilder.com. The job ads distribution at the SOC major level is skewed towards occupation categories such as 15 (Computer and Mathematical) and 41 (Sales) among others. Job ads from occupation categories such as 45 (Farming, Fishing, and Forestry) and 23 (Legal) comprise the long tail of the overall job ads distribution. We tackle the class imbalance problem in the dataset by under sampling all the classes to a base count of around 150 k jobs. This reduced the training dataset size to 2 m jobs.

Since crowdsourced labeling of datasets is susceptible to systematic and consistency errors, we used Autocoder[1] to label the training dataset. Autocoder is a third-party tool that classifies recruitment industry content such as resumes and job ads to O*NET codes. Autocoder's classification accuracy of 80 % for job titles and 85 % for job ads puts a theoretical upper bound on Carotene's potential accuracy. However, we proceeded with using Autocoder labels for our dataset because one of our long-term goals is to meet or exceed Autocoder's accuracy and because of the cost and efficiency gains realized compared to third-party crowdsourcing alternatives. Figure 6.2 gives an overview of Carotene.

---

[1]http://www.onetsocautocoder.com/plus/onetmatch?action=guide

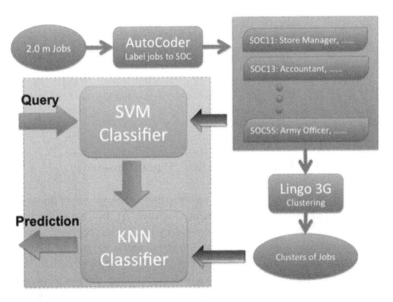

Fig. 6.2 Architecture of Carotene

6.1.2.1   Taxonomy Discovery Using Clustering

The first phase of Carotene uses a clustering process to discover job titles. The training dataset is organized by SOC major into dataset segments and preprocessed by removing extraneous markup characters and noise. For clustering, we use Lingo3G [7], a proprietary clustering algorithm. Lingo3G identifies meaningful cluster label that assists in inferring a job title taxonomy from a dataset. Some of the Lingo3G parameter settings that we empirically set were *max-cluster-size*, *min-cluster-size*, and *merge-threshold* among others. We run Lingo3G on these dataset segments to create *verticals*. These verticals are clusters of job titles for a SOC major. Every cluster in a vertical has a *glabel* that is the representative normalized label of the raw job titles in that cluster. We also applied several post-processing steps such as glabel validation and cluster merging and deletion. These steps require domain expertise and human validation. To facilitate such post-processing operations, we use an in-house tool that provides a user-friendly interface and abstraction to automate most of these tasks. The tool helped us reduce our original estimate of 3 months for creating the taxonomy to 1 month. The fine-level classifier in the classification component of Carotene uses the appropriate vertical to return the classification for the input query text. There are 4,076 job titles in the Carotene taxonomy. Figure 6.3 shows a Physical Therapy Assistant cluster from the SOC-31 vertical.

```
22 | Physical Therapy Assistant (871)
   |   *Physical Therapist Assistant - Orlando - Acute Care - Per Diem
   |   *Physical Therapist Asst. - STE
   |   *Physical Therapy Assistant - Deaconess Hospital
   |   *Physical Therapy Assistant - Hillside Rehabilitation Hospital
   |   *Physical Therapy Assistant- Per Diem- Northside Medical Center
```

Fig. 6.3 Physical therapy assistant cluster

6.1.2.2   Coarse-Level Classification: SOC Major Classifier

Carotenes' classification component is a coarse and fine-level hierarchical cascade. Our coarse-level classifier is a SOC major classifier. At the coarse level, jobs are classified to a SOC major which is then used by the fine-level classifier to narrow the space of candidate job title labels for the input query. We use the LIBLINEAR [8] implementation of the SVM classifier as the SOC major classifier. We chose the L2-regularized L2-loss SVC (dual) solver for training the model because this is a large-scale, sparse data-learning problem where the number of features is far greater than the number of training instances. The training dataset had 5.8 m features with approximately 280 nonzero features per instance. It took 1.5 h to train the SOC major classifier on an m3.2xlarge EC2 instance with 30GB of RAM. While the SOC major classifier is a multi-class, multi-label classifier that returns a list of classifications ranked by confidence, in Carotenes' current cascade architecture, only the top-ranked label is sent to the fine-level classifier for further processing. We estimated the accuracy of the SOC major classifier using tenfold cross validation where the average precision, recall, and F-scores were 95.54 %, 95.33 %, and 95.43 %, respectively. Global accuracy was 95.8 % and coverage was 99 %.

6.1.2.3   Fine-Level Classification: Proximity-Based Classifier

Proximity-based classifiers such as k-nearest neighbor (kNN) find the k-nearest neighbors of a query instance or pre-process the data into clusters and use cluster meta-data as k-nearest neighbor documents to improve classification accuracy. Carotene uses kNN as a multi-class, multi-label proximity-based fine-level classifier with k empirically set to 20. Standard implementations of kNN are considered slower than other classification techniques because of kNNs' lazy learning characteristics. We use a Lucene[2]-based implementation of kNN that results in classification response time of less than 100 ms. The kNN classifier has access to 23 verticals and chooses the appropriate vertical based on the classification it receives from the SOC major classifier. The first version of Carotene, CaroteneV1, had a classification component that was a single-level kNN classifier that queried a

---

[2]http://lucene.apache.org

single vertical composed of the initial set of 1,800 titles. Compared to its majority voting-based implementation in CaroteneV1 which is susceptible to skewed class distributions, the kNN classifier in Carotene has an improved voting strategy where the absolute values of Lucene scores are used to assign a weight to each neighbor. In CaroteneV1, only the raw job titles in verticals were indexed and searched on. Raw job titles are noisier than normalized titles. In Carotene, the verticals also have a normalized job title called glabel for each cluster; hence, both the raw titles and glabels are indexed as *title* and *label* fields, respectively. Carotene uses a Lucene multi-field query with a default boost factor of 1.0 and a boost factor of 0.9 for the title and label fields, respectively. Indexing the glabel improves the quality of classification in close boundary decisions.

6.1.3   Experimental Results and Discussion

Development teams at CB use Carotene for classifying large text such as job ads and resumes as well as short text consisting of job titles. Hence, we used two datasets to compare the performance of Carotene to CaroteneV1: job ads and normalized titles. The job ads dataset contains 1,000 randomly sampled jobs posted on CareerBuilder.com. We ensured that the job ads selected were from all the 23 major SOC categories. For the normalized title dataset, we chose 570 job titles that exist in both CaroteneV1 and Carotene taxonomies. Table 6.2 shows the performance of CaroteneV1 and Carotene on the two datasets. We observe a 27.7 % increase in accuracy at classifying job ads with Carotene compared to CaroteneV1. For the job title dataset, there is a 12 % increase in accuracy from 68 % to 80 %. The more comprehensive taxonomy along with the cascade architecture of the classification component of Carotene gives better job title classification results.

We also compared Carotene to Autocoder by running a user survey of users registered on CareerBuilder.com. We used surveys as a way of measuring their comparative performances because the two systems have different taxonomies. The most recent work history listed in a user's resume was classified using both Carotene and Autocoder, and the users were then requested to rate the classifications on a scale of 1–5, where 5 is for a perfect classification and 1 for a classification that totally missed the mark. We surveyed 2 M active users, and Carotene had a classification precision of 68 % versus 64 % for Autocoder. We observed that in some instances, Autocoder's classification was more likely to be given a higher rating because of the more general nature of its classifications especially

Table 6.2 Accuracy of Carotene and CaroteneV1 on job ads and job titles		Accuracy	
	Dataset	Carotene V1 (%)	Carotene (%)
	Normalized titles	68	80
	Job ads	39.7	67.4

when the more fine-grained Carotene classification did not exactly match the user's job title. For example, for "Embedded Systems Software Engineer," the Autocoder classification "Software Developers, Systems Software" is more likely to be rated higher by users than the Carotene classification of "C++ Developer." As we iteratively improve Carotene's classification accuracy as well as expand the coverage of its taxonomy, we expect such discrepancies in user ratings to decrease over time.

As a system currently in production at CB and actively being used by several internal clients, there are several improvements and extensions to Carotene that we are working on such as (1) internationalization, support for Carotene in international markets, (2) extending the architecture to support classification at the O*NET code level, and (3) enrichments vectors and semantic kernels for more semantically aligned classifications.

6.2   CARBi: A Data Science Ecosystem

CARBi is a Big Data/Data Science ecosystem designed and implemented to satisfy the growing needs of data scientists at CB. With CARBi, data scientists have support for developing, executing, and managing Big Data solutions. The core components of CARBi are explained in the following subsections.

6.2.1   Accessing CB Data and Services Using WebScalding

WebScalding was initially developed to process high volumes of resume and job data (most of our data processing are implemented as scalable webservices, hence the name "WebScalding"). The project evolved to an umbrella of many reusable components and encompasses several development best practices. Today, WebScalding is a collective name for a group of projects that are designed for Big Data solutions. A project can be a WebScalding project or a WebScalding-compatible project, and each of them is explained in the following subsections:

- *WebScalding project*: A project that uses the WebScalding core library that is a Big Data library developed using Twitter's Cascading [9] interface called Scalding.[3] In addition to the core project, a WebScalding project can use any WebScalding compatible projects as well.
- *WebScalding-compatible project*: Any project that exposes APIs satisfying two main requirements: (1) independent execution, each execution of the API should be completely independent of any another execution thus enabling parallel execution of the API, and (2) independent existence, every execution is self-contained

---

[3]https://github.com/twitter/scalding

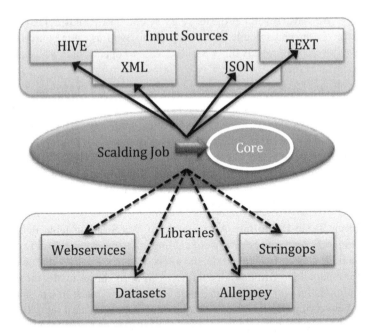

Fig. 6.4 Overview of WebScalding

such that it can create data if required. As an example, if an API uses an index, the WebScalding-compatible implementation of the API can create the index if the index doesn't exist.

An overview of the WebScalding system is shown in Fig. 6.4. Currently, the system can read/write data from/to four formats: (1) HIVE, (2) XML, (3) TEXT, and JSON. In the figure, four WebScalding-compatible projects are shown and are explained in the following subsections:

1. *Web Services*: We designed the web services component to enable us to call our scalable web services for high volumes of data. Using the web services component, users can define a web service as an executable Hadoop job reading inputs from any of the supported formats and writing the output to one of the supported format.
2. *StringOps*: Our primary sources of data are job postings and resumes. Consequently, text/string processing is an important functionality for WebScalding applications. The StringOps component supports many text operations such as stemming, string matching, string sorting, language detection, and parsing several text formats.
3. *Alleppey*: Our commonly used machine learning libraries and tools are available in the Alleppey project. This project enables us to quickly try out different machine learning models for the same data. We currently support several popular machine learning toolkits (e.g., WEKA [10], LibShortText [11]).

4. *Datasets*: We have several projects (e.g., sCooL [12], SKILL [13]) that use knowledge bases (KBs) built using our data as well as open source datasets. We have developed projects that provide access to open source datasets such as Wikipedia and Freebase.[4]

To illustrate how WebScalding can help data scientists in designing and executing Big Data solutions, we have included an example to show how HIVE tables are accessed in a WebScalding project. For WebScalding, any HIVE table is a file location, so it is possible to have multiple WebScalding table definitions for the same file location. This is very useful for cases where the actual HIVE table has many columns, but the user is interested in only some of them for a given job. To access any HIVE table in a WebScalding project, we have to first define a WebScalding table and then load that WebScalding table in a Hadoop job.

WebScalding Table Definition An example of WebScalding table definition is shown in Fig. 6.5. We require every WebScalding job to be executable in two modes: (1) local and (2) cluster. In local mode, the job is executed locally with local files as the input and output are written to local files as well. In cluster mode, the job reads from Hadoop HDFS and writes to the HDFS as well. To make this possible for a Hadoop job involving HIVE tables, we require the users to create a sample table that can simulate the HIVE table. In the example in Fig. 6.5, our goal is to read a table

```
1    case class UnifiedProfile(edgeId: String, profiles: String,
2                              parentId: String)
3
4    object UnifiedProfileTable extends HiveTable[UnifiedProfile] {
5
6      def tableName: TableName = TableName("local.txt", "remote")
7
8      def localColumns(x: String) = {
9        val parts = x.split("\t", -1)
10
11       UnifiedProfile(parts(0).trim, parts(4).trim(), parts(13).trim)
12     }
13
14     def hdfsColumns(x: String) = {
15       val parts = x.split(HIVEDelimiter, -1)
16       UnifiedProfile(parts(0).trim, parts(4).trim(), parts(13).trim)
17
18     }
19
20     def columns = Columns(localColumns, hdfsColumns)
21
22   }
```

Fig. 6.5 HIVE table in WebScalding

---

[4]Freebase, http://www.freebase.com

```
1 class GenerateEdgeNetworkLinks(args: Args)
2                        extends Job(args) with HiveAccess  {
3
4    getHiveContent(UnifiedProfileTable)
5 }
```

Fig. 6.6 Using HIVE table in a Scalding job

and create UnifiedProfile object (line 1) for each row. Users have to provide the information how to convert a line in the file location to a UnifiedProfile object. From our experience, this parsing can be different (e.g., the delimiters can be different, number of columns can be different) for cluster and local modes; hence, users has the flexibility to provide the conversion process separately for local (lines 8–12) and remote (lines 14–18) modes.

WebScaling Table Access Figure 6.6 shows an example of accessing a WebScalding Table in a Hadoop job. A WebScalding table can be accessed in a Hadoop job using the trait HiveAccess (line 2) which includes a method getHiveContent(WebScaldingTable) (line 4). This returns a TypedPipe[5] of type UnifiedProfile.

To summarize, we achieved two main goals by introducing WebScalding: (1) we raised the level of abstraction, enabling data scientists to focus more on the problem at hand than on the low level implementation details of MapReduce or Hadoop, and (2) we empowered users to develop "write once and run on local and cluster modes" code, enabling users to test the data flow locally before executing it on a cluster.

6.2.2 ScriptDB: Managing Hadoop Jobs

The main motivation of ScriptDB is management of multiple Hadoop jobs associated with a single project. Some of these jobs are interconnected to form a pipeline while others stay independent (e.g., the Recruitment Edge project[6] currently has more than 100 Hadoop jobs). Such Hadoop jobs usually require different configuration parameters, input arguments, and other details that are very specific to the project and the job in question. In an effort to keep all the Hadoop jobs available to run, we have the tedious task of maintaining script files specific to each Hadoop job. Any refactoring in the project or the job can lead to making changes in these script files. Moreover, changing versions or platforms (e.g., Hadoop versions, Hadoop to Scalding conversions) can cause significant changes in the script files for each Hadoop job execution.

---

[5] TypedPipe, http://twitter.github.io/scalding/index.html#com.twitter.scalding.typed.TypedPipe
[6] Recruitment Edge, http://edge.careerbuilder.com

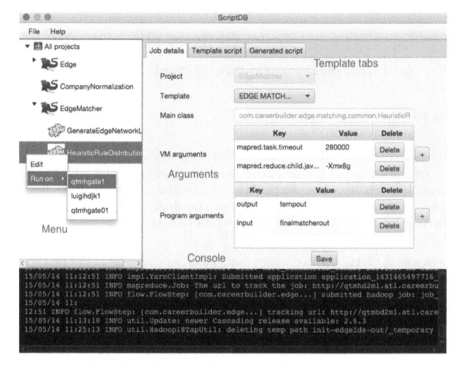

Fig. 6.7 ScriptDB app to execute Hadoop jobs

We developed ScriptDB so that data scientists can (1) browse through the available Hadoop jobs and if required submit the job to a given cluster and (2) add/update a Hadoop job independent of the language or platform. A screenshot of ScriptDB is shown in Fig. 6.7. To define a Hadoop job using ScriptDB, the user has to specify the project, a script template, the main class for the Hadoop job, and arguments. The key parts of the ScriptDB are explained in the following subsections:

- *Menu*: Hadoop jobs are organized by projects. This is to facilitate an ordering and also to allow project-specific templates for job submission.
- *Arguments*: Using ScriptDB, users can provide two kinds of arguments: (1) program arguments, the arguments the Scalding job/Hadoop job need to run the job, and (2) virtual machine (VM) arguments, similar to VM arguments for a JAVA program, the configuration parameters for a MapReduce job (e.g., mapred.task.timeout).
- *Console*: Console prints out the information messages and error messages for the user. This is also useful if user wants to execute a command in a server, in that case, console serves as "System.out."
- *Template tabs*: The key feature of the ScriptDB is the template. The template mechanism makes ScriptDB language independent. The user is given the ability to create/update templates and to view the generated script file from the template for a given job.

```
template1(main,arguments,jarname,options,projectname,projectlocation) ::=<<
#!/bin/bash
JAR_NAME=<projectlocation>/<jarname>
RUNNER=com.careerbuilder.edge.main.JobRunner
MAIN_CLASS=<mainclass>
cd <projectlocation>
git pull origin yarn-changes
sbt 'project matcher' clean assembly
chmod a+r <jarname>
ARGUMENTS=<arguments:{u|--<u.key> <u.value> }>
hadoop jar $JAR_NAME  $RUNNER $MAIN_CLASS --hdfs $ARGUMENTS &
```

Fig. 6.8 Stringtemplate script for executing a Hadoop job

```
1 #!/bin/bash
2 JAR_NAME=/home/fjacob.site/Recruitment-Edge/target/scala-2.11/EdgeHadoop-job.jar
3 RUNNER=com.careerbuilder.edge.main.JobRunner
4 MAIN_CLASS=com.careerbuilder.edge.emsi.GenerateEdgeNetworkLinks
5 cd /home/fjacob.site/Recruitment-Edge/scalding/Edge
6 git pull origin yarn-changes
7 sbt 'project matcher' clean assembly
8 chmod a+r matcher/target/scala-2.11/EdgeHadoop-job.jar
9 ARGUMENTS=--output init-edgeIds-out --input init-edgeIds
10 hadoop jar $JAR_NAME  $RUNNER $MAIN_CLASS --hdfs $ARGUMENTS &
```

Fig. 6.9 Code generated using the ScriptDB template engine

ScriptDB uses Stringtemplate[7] as the template engine, and hence, the templates have to be specified using the Stringtemplate syntax. In Fig. 6.8, all the StringTemplate-specific syntax is highlighted in blue. An example code generated using the template is shown in Fig. 6.9.

References

1. Dumais S, Chen H (2000) Hierarchical classification of web content. In: Proceedings of ACM SIGIR'00. New York, USA, pp 256–263
2. Joachims T (1999) Transductive inference for text classification using support vector machines. In: Proceedings of ICML 1999. San Francisco, USA, pp 200–209
3. Liu M, Lu L, Ye X et al (2011) Coarse-to-fine classification via parametric and nonparametric models for computer-aided diagnosis. In: Proceedings of ACM CIKM'11. New York, USA, pp 2509–2512
4. Shen D, Ruvini J-D, Sarwar B (2012) Large-scale item categorization for e-commerce. In: Proceedings of ACM CIKM'12, pp 595–604

---

[7]Stringtemplate, http://www.stringtemplate.org

5. Bekkerman R, Gavish M (2011) High-precision phrase-based document classification on a modern scale. In: Proceedings of the 17th ACM SIGKDD-KDD'11, pp 231–239

6. Babbar R, Partalas I (2013) On flat versus hierarchical classification in large-scale taxonomies. In: Proceedings of the neural information processing systems (NIPS), pp 1–9

7. Osiński S, Weiss D (2005) A concept-driven algorithm for clustering search results. IEEE Intell Syst 3(20):48–54

8. Fan RE et al (2008) Liblinear: a library for large linear classification. J Mach Learn Res 9:1871–1874

9. Nathan P (2013) Enterprise data workflows with cascading, 1st edn. O'ReillyMedia, Sebastopol. Sebastopol, USA

10. Hall M, Frank E, Holmes G et al (2009) The WEKA data mining software: an update. ACM SIGKDD Explor Newsl 11(1):10–18

11. Yu H-F, Ho C-H, Juan Y-C et al (2013) LibShortText: a library for short-text classification and analysis. Department of Computer Science, National Taiwan University, Taipei, Technical report. http://www.csie.ntu.edu.tw/ncjlin/papers/libshorttext.pdf

12. Jacob F, Javed F, Zhao M et al (2014) sCooL: a system for academic institution name normalization. In: 2014 international conference on collaboration technologies and systems, CTS. Minneapolis, USA, pp 86–93

13. Zhao M, Javed F, Jacob F et al (2015) SKILL: a system for skill identification and normalization. AAAI 2015. Austin, USA, pp 4012–4018

Chapter 7
Extraction of Bayesian Networks from Large Unstructured Datasets

Marcello Trovati

Abstract Bayesian networks (BNs) provide a useful modelling tool with a wide applicability on a variety of research and business areas. However, their construction is very time-consuming when carried out manually. In this chapter, we discuss an automated method to identify, assess and aggregate relevant information from large unstructured datasets to build fragments of BNs.

7.1 Introduction

Bayesian networks (BNs) are graphical models that capture independence relationships among random variables, based on a basic law of probability called *Bayes' rule* [1]. They are a popular modelling framework in risk and decision analysis, having been used in a variety of applications such as safety assessment of nuclear power plants, risk evaluation of a supply chain and medical decision support tools [2–5]. More specifically, BNs consist of a graph, whose nodes represent objects based on a level of uncertainty, also called *random variables*, and whose edges indicate a dependence relationship between them. Beside this graphical representation, Bayesian networks also contain quantitative information which represents a factorisation of the joint probability distribution of all the variables in the network. In fact, each node has an associated conditional probability table which captures the probability distribution associated with the node conditional on each possibility. Suppose, for example, we want to explore the chance of finding wet grass on any given day. In particular, assume the following:

1. A cloudy sky is associated with a higher chance of rain.
2. A cloudy sky affects whether the sprinkler system is triggered.
3. Both the sprinkler system and rain have an effect on the chance of finding wet grass.

M. Trovati (✉)
Department of Computing and Mathematics, University of Derby, Derby, UK
e-mail: M.Trovati@derby.ac.uk

© Springer International Publishing Switzerland 2015　　　　　　　　　　97
M. Trovati et al. (eds.), *Big-Data Analytics and Cloud Computing*,
DOI 10.1007/978-3-319-25313-8_7

Fig. 7.1 An example of a BN

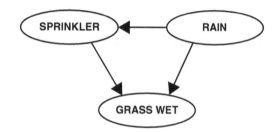

In this particular example, no probabilistic information is given. The resulting BN is depicted in Fig. 7.1.

It is clear that such graphical representation provides an intuitive way to depict the dependence relations between variables.

In the definition of BNs, the most complex statements do not refer to dependencies, but rather about independences (i.e. absence of edges in the graph), as it is always possible to determine dependence through the conditional probability tables when an edge is present, even though the reverse is not true.

The construction of a BN can be done either through explicit data, or when not readily available, via literature review or expert elicitation. While the first approach can be automated, the other ones necessitate a significant amount of manual work which can result impractical on a large scale. There is extensive research on Bayesian networks, and in particular their extraction from text corpora has been attracting considerable attention. For example, in [6] the authors suggest a domain-independent method for acquiring text causal knowledge to generate Bayesian networks. Their approach is based on a classification of lexico-syntactic patterns which refer to causation, where an automatic detection of causal patterns and semi-validation of their ambiguity is carried out. Similarly, in [7] a supervised method for the detection and extraction of causal relations from open domain texts is presented. The authors provide an in-depth analysis of verbs, cue phrases that encode causality and, to a lesser extent, influence.

7.2 Text Mining

Text mining (TM) consists of a range of computational techniques to achieve human language understanding by using linguistic analysis for a range of tasks or applications. In particular, such methods have been shown to be crucially important in the way we can represent knowledge described by the interactions between computers and human (natural) languages. The main goal of TM is to extend its methods to incorporate any language, mode or genre used by humans to interact with one another in order to achieve a better understanding of the information captured by human communication [8].

Language is based on grammatical and syntactic rules which can be captured by *patterns* or, in other words, templates that sentences with similar structure follow. Such language formats enable the construction of complex sentences, as well as framing of the complexity of language. For example, the ability to identify the subject and the object of a particular action described by a sentence has heavily contributed to human evolution and the flourishing of all civilisations.

7.2.1 Text Mining Techniques

TM methods have been developed to embrace a variety of techniques which can be grouped into four categories: *symbolic, statistical, connectionist and hybrid*. In this section, we will briefly discuss these approaches in terms of their main features and suitability with respect to the tasks requiring implementation.

- *Symbolic approach*. In this method, linguistic phenomena are investigated which consist of explicit representation of facts about language via precise and well-understood knowledge representation [9]. Symbolic systems evolve from human-developed rules and lexicons which produce the relevant information basin this approach is based on. Once the rules have been fully defined, a document is analysed to pinpoint the exact conditions which validates them. All such rules associated with semantic objects generate networks describing their hierarchical structure. In fact highly associated concepts exhibit directly linked properties, whereas moderately or weakly related concepts are linked through other semantic objects. Symbolic methods have been widely exploited in a variety of research contexts such as information extraction, text categorisation, ambiguity resolution, explanation-based learning, decision trees and conceptual clustering.
- *Statistical approach*. Observable data and the investigation of large documents are used to develop generalised models based on smaller knowledge datasets and significant linguistic or world knowledge [8]. The set of states are associated with probabilities, and there are several techniques that can be used to investigate their structures such as the hidden Markov model where the set of status is regarded as not directly observable. They have many applications such as parsing rule analysis, statistical grammar learning and statistical machine translation to name but a few.
- *Connectionist approach*. This method integrates statistical learning with representation techniques to allow an integration of statistical tools with logic-based rule manipulation, generating a network of interconnected simple processing units (often associated with concepts) with edge weights representing knowledge. This typically creates a rich system with an interesting dynamical global behaviour induced by the semantic propagation rules. Troussov et al. [10] investigated a connectionist distributed model pointing towards a dynamical generalisation of syntactic parsing, limited domain translation tasks and associative retrieval.

7.2.2 General Architecture and Various Components of Text Mining

In Linguistics, a (formal) grammar is a set of well-defined rules which govern how words and sentences are combined according to a specific syntax. A grammar does not describe the meaning of a set of words or sentences, as it only addresses the construction of sentences according to the syntactic structure of words. Semantics, on the other hand, refers to the meaning of a sentence [8]. In computational linguistics, semantic analysis is a much more complex task since its aim is the full understanding of the meaning of text.

Any text mining process consists of a number of steps to identify and classify sentences according to specific patterns, in order to analyse a textual source. Broadly speaking, in order to achieve this, we need to follow these general steps:

1. Textual data sources are divided into small components, usually words, which can be subsequently syntactically analysed.
2. These, in turn, create *tokenised* text fragments, which are analysed according to the rules of a formal grammar. The output is a parsing tree or, in other words, an ordered tree representing the hierarchical syntactic structure of a sentence.
3. Once we have isolated the syntactic structure of a text fragment, we are in the position of extracting relevant information, such as specific relationships, sentiment analysis, etc.

More specifically, the main components of text mining are as follows (Fig. 7.2):

7.2.3 Lexical Analysis

Lexical analysis is the process which analyses the basic components of texts and groups into tokens [11]. In other words, lexical analysis techniques identify the syntactic role of individual words which are assigned to a single part-of-speech tag.

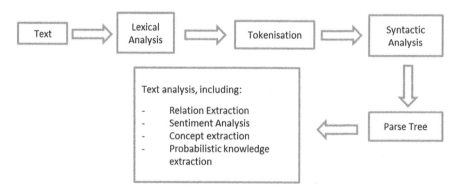

Fig. 7.2 The main components of text mining

Lexical analysis may require a lexicon which is usually determined by the particular approach used in a suitably defined TM system, as well as the nature and extent of information inherent to the lexicon. Mainly, lexicons may vary in terms of their complexity as they can contain information on the semantic information related to a word. Moreover, accurate and comprehensive subcategorisation lexicons are extremely important for the development of parsing technology as well as crucial for any NLP application which relies on the structure of information related to predicate-argument structure. More research is currently being carried out to provide better tools in analysing words in semantic contexts (see [12] for an overview).

7.2.4 Part-of-Speech Tagging

Part-of speech tagging (POS) allows to attach a specific syntactic definition (noun, verb, adjective, etc.) to the words which are part of a sentence. This task tends to be relatively accurate, as it relies on a set of rules which are usually unambiguously defined. Often POS tasks are carried out via the statistical properties of the different syntactic roles of tokens [8]. Consider the word *book*. Depending on the sentence it belongs to, it might be a verb or a noun. Consider "a book on chair" and "I will book a table at the restaurant". The presence of specific keywords, such as "a" in the former and "I will" in the latter, provides important clues as to the syntactic role that *book* has in the two sentences. One of the main reasons for the overall accuracy of POS tagging is that a semantic analysis is often not required, as it is based on the position of the corresponding token.

7.2.5 Parsing

Once the POS tagging of a sentence has identified the syntactic roles of each token, each sentence can be considered in its entirety. The main difference with POS tagging is the fact that parsing enables the identification of the hierarchical syntactic structure of a sentence. Consider, for example, Fig. 7.3 which depicts the parsing tree structure of the sentence "This is a parsing tree". Note that each word is associated with a POS symbol which corresponds to its syntactic role [8].

An important aspect of text analysis is the ability to determine the type of the entities, which refer to words, or collections of them. For example, determining whether a noun refers to a person, an organisation or geographical location substantially contributes to the extraction of accurate information and provides the tools for a deeper understanding. For example, the analysis of "dogs and cats are the most popular pets in the UK" would identify that dogs and cats are animals and the UK is a country. Clearly, there are many instances where this depends on the context. Think of "India lives in Manchester". Anyone reading such sentence would interpret, and rightly so, India as the name of a specific person. However, a computer

Fig. 7.3 The parsing tree of the sentence "This is a parsing tree"

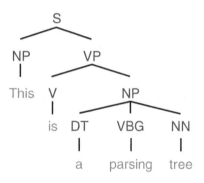

might not be able to do so and determine that it is a country. We know that a country would not be able to "live" in a city. It is just common sense. Unfortunately, computers do not have the ability to discern what common sense is. They might be able to guess according to the structure of a sentence or the presence of specific keywords. This is a very effective example of semantic understanding, which comes natural to humans but a very complex task to computers.

7.2.6 Named Entity Recognition

Coreference resolution is the process of determining which text components refer to the same objects. For example, *relation resolution* attempts to identify which individual entities or objects a relation refers to. Consider the following sentence, "We are looking for a fault in the system". Here, we are not looking for *any* fault in the system, rather for a specific instance.

7.2.7 Named Entity Recognition

The identification of relations between different entities within a text provides useful information that can be used to determine quantitative and qualitative information linking such entities. For example, consider the sentence "smoking potentially causes lung cancer". Here, the act of smoking is linked to lung cancer by a causal relationship. This is clearly a crucial step in building BNs, even though such analysis requires a deep understanding of the associated textual information.

7.2.8 Concept Extraction

A crucial task in information extraction from textual sources is concept identification, which is typically defined as a one or more keywords or textual definitions.

The two main approaches in this task are supervised and unsupervised concept identification, depending on the level of human intervention in the system.

In particular, formal concept analysis (FCA) provides a tool to facilitate the identification of key concepts relevant to a specific topical area [5, 13]. Broadly speaking, unstructured textual datasets are analysed to isolate clusters of terms and definitions referring to the same concepts, which can be grouped together. One of the main properties of FCA allows user interaction, so that user(s) can actively operate the system to determine the most appropriate parameters and starting points of such classification.

7.2.9 Sentiment Analysis

Semantic analysis determines the possible meanings of a sentence by investigating the interactions among word-level meanings in the sentence. This approach can also incorporate the semantic disambiguation of words with multiple senses. Semantic disambiguation allows the selection of the sense of ambiguous words, so that they can be included in the appropriate semantic representation of the sentence [14]. This is particularly relevant in any information retrieval and processing system based on ambiguous and partially known knowledge. Disambiguation techniques usually require specific information on the frequency with which each sense occurs in a particular document, as well as on the analysis of the local context, and the use of pragmatic knowledge of the domain of the document. An interesting aspect of this research field is concerned with the purposeful use of language where the utilisation of a context within the text is exploited to explain how extra meaning is part of some documents without actually being constructed in them. Clearly this is still being developed as it requires an incredibly wide knowledge dealing with intentions, plans and objectives [8]. Extremely useful applications in TM can be seen in inference techniques where extra information derived from a wider context successfully addresses statistical properties [15].

7.3 The Automatic Extraction of Bayesian Networks from Text

As discussed above, due to the mathematical constraints posed by Bayes' rule and general probability theory, the identification of suitable Bayesian networks is often carried out by human intervention, in the form of a modeller who identifies the relevant information. However, this can be extremely time-consuming and based on only specific, often limited, sources depending on the modeller's expertise. As a consequence, the ability to automatically extract the relevant data would potentially add enormous value in terms of increased efficiency and scalability to the process of defining and populating BNs. However, extracting both explicit and implicit information, and making sense of partial or contradictory data, can be a complex challenge. More specifically, elements of the quantitative layer depend on the graphical layer, or in other words the structure of the conditional probability tables depends on the parents of each node. Therefore, it is necessary to determine the structure of the BN before populating it with quantitative information.

7.3.1 Dependence Relation Extraction from Text

Nodes in BNs, which are connected by edges, indicate that the corresponding random variables are dependent. Such dependence relations must be therefore extracted from textual information, when present. The conditional dependencies in a Bayesian Network are often based on known statistical and computational techniques, which are based on a combination of methods from graph theory, probability theory, computer science and statistics. Linguistically speaking, a dependence relation contains specific keywords which describe that two concepts are related to a certain degree. Consider the sentence "lung cancer is more common among smokers". There is little doubt that we would interpret this as clear relation linking lung cancer with smoking. However, there is not a precise linguistic definition to determine a relationship between two concepts from text, due to its content dependence. When a full automation of the process of textual information extraction is carried out, a clear and unambiguous set of rules ensures a reasonably good level of accuracy. As a consequence, it is usually advisable to consider *causal relationships*, which are a subgroup of dependence relationships [1]. In fact, they are likely to convey a much stronger statement, and they are more easily identified due to a more limited set of linguistic rules that characterise them. Going back to the above example, saying that smoking *causes* lung cancer assumes a direct link between them. We cannot arguably say the contrary, but there are other cases where there is a less marked cut-off. If we are only looking for causal relationships when populating a BN, we might miss out several dependence relations. However, accuracy is much more preferable. The integration of an automatic BN extraction with human intervention usually addresses such issue.

Causal learning often focuses on long-run predictions through an estimation of the parameters of a causal Bayes network structural learning. An interesting approach is described in [16] where people's short-term behaviour is modelled through a dynamical version of the current approaches. Moreover the limitation of a merely static investigation is addressed by a dynamical approach based on BNs methods. Their result only applies to a particular scenario, but it offers a new perspective and it shows huge research potential in this area.

7.3.2 Variable Identification

Mapping a representative to a specific variable is closely linked to the task of relations extraction. However, this is partially a modelling choice by the user based on the set of relevant concepts. Consider again the sentence "smoking causes lung cancer". If this was rephrased as "smokers are more likely to develop lung cancer", we would need to ensure that "smoking" and "smokers" are identified as a single variable associated with the act of smoking. In a variety of cases, this can be addressed by considering synonymy. However, such as in our example, it might also happen that they refer to the same concept, rather than being the *same* concept. Formal concept analysis (FCA) is one of the computational techniques that can be successfully applied in this particular context 5.

7.3.3 BN Structure Definition

This step performs the aggregation of the information gathered in the previous two steps to output the structure of the BN. This includes the evaluation of the credibility of each dependence relation, determining whether the dependence stated is direct or indirect and ensuring that no cycles are created in the process of orienting the edges.

7.3.4 Probability Information Extraction

An essential part in the extraction and subsequent creation of BNs involves the processing of the textual sources to determine probability of variables. Consider the following examples:

1. If you press this button, that door will open almost certainly.
2. A thunderstorm is triggered if the overall humidity exceeds 80 %.
3. Smoking increases the chances of getting lung cancer.
4. Non-smokers are less likely to die of breast cancer.
5. Considering that alcoholism is heavily influenced by a propensity to addictive behaviour, it is not reasonable to assume that liver damage is independent of alcoholism.

All of these sentences capture some probabilistic relationships between concepts. However, none of them provide conclusive and unambiguous information, which can be utilised to reason. In fact, the combination of qualitative and quantitative data creates a variety of challenges, which need to be addressed to produce relevant and accurate information.

Consider, for example, (1)–(4). None of them give a direct and uniquely interpretable insight into their specific probabilistic information, as they all depend on how specific keywords are interpreted. A relatively naïve solution is the identification, assessment and ranking of such keywords (and their combinations) when describing probability. There are, however, a variety of issues. How can we uniquely rank such keywords, since potentially a huge combination can create different nuances of meaning? For example, "quite probably" is certainly not the same as "likely". However, is "slightly improbable" similar to "not very certain"? The definition of a single measure to address such issue would simply depend on the modeller's choice.

Furthermore, the underlying objective of big data depends on the interrelations of diverse and multidisciplinary topics, resulting in the intrinsic difficulty in finding a common ground in terms of the linguistic features that specific probabilistic description should have. Furthermore, the automatic extraction from text is even more penalised by such vagueness, as any disambiguation techniques would be negatively affected by such scenario.

In Trovati and Bessis [17], an automated method to assess the influence among concepts in unstructured sets is introduced. Despite not being directly related to BNs, it shows potential in the extraction of the mutual co-occurrence properties between nodes.

7.3.5 Probability Information Extraction

This step integrates the steps discussed above to construct fragments of BNs via user's interaction.

Figure 7.4 depicts this process in a sequential set of steps. However, the repeated implementation of such steps in a series of loops might be required to obtain meaningful BNs fragments.

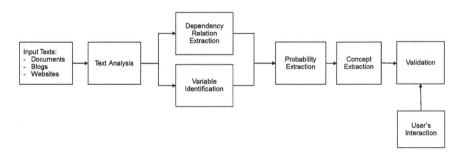

Fig. 7.4 The architecture of Bayesian network extraction from textual information

7.3.6 General Architecture

The general architecture of the extraction of fragments of BNs from text corpora consists of the following components:

1. Existing and predefined information on specific topics would be incorporated into a database or *knowledge database* (KDB) consisting of:

    (a) Historical data from structured DBs
    (b) Bayesian networks built on existing data
    (c) Data entered by modeller(s) and manually validated

    The KDB is an important component since it is based on information which is considered "reliable". In a variety of cases, the KDB is maintained by modelling experts to ensure that the data is regularly updated to prevent any inconsistency and ambiguity.

2. The user would interact with the system by specifying further textual sources and structured datasets.

3. The extraction and data aggregation stage consists of the identification of the appropriate textual data associated with such unstructured datasets, as well as the removal of any data duplication. An essential part of this process is to address any qualitatively and quantitatively inconsistency. As discussed above, BNs have strict mathematical constraints which make any fully unsupervised automatic extraction prone to inaccuracies and inconsistencies. As a consequence, human intervention is often advisable to minimise any such error.

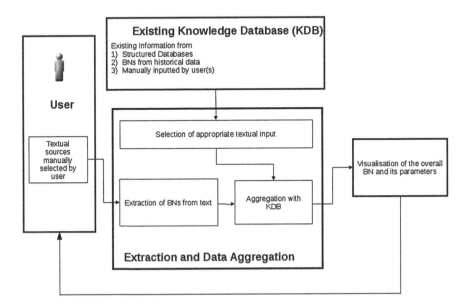

Fig. 7.5 General architecture of Bayesian networks

4. Finally, the BN is visualised, providing:

   (a) Relevant information on the structure of the BN
   (b) Description of the different parameters
   (c) Any required action in order to address any inconsistency which could not be resolved automatically. This is typically an interactive step, where the result can be updated by the user as well as focused on a specific part of the BN (Fig. 7.5).

7.4 Conclusions

In this chapter, we have discussed the creation of fragments of BNs from text corpora, whose applicability has enormous advantages in a multitude of areas. In particular, big data research would benefit from the use of an automated extraction of BNs to enable the extraction of relevant information to inform and facilitate the decision-making process based on datasets, which are often difficult to manually manipulate and assess.

References

1. Pearl J (1998) Probabilistic reasoning in intelligent systems: networks of plausible inference. Morgan Kaufmann Publishers, San Francisco
2. Ghazi A, Laskey K, Wang X et al (2006) Detecting threatening behavior using Bayesian networks. C4I papers
3. Chen, H (2008) Homeland security data mining using social network analysis. In: IEEE international conference on intelligence and security informatics. Taipei
4. McKenna JA (2004) The internet and social life. Annu Rev Psychol 55:573–590
5. Poelmans JE (2010) Formal concept analysis in knowledge discovery: a survey. In: International conference on conceptual structures (ICCS). Kuching, Sarawak, Malaysia
6. Sanchez-Graillet O, Poesio M (2004) Acquiring Bayesian networks from text. In: LREC
7. Liddy ED (2001) A robust risk minimization based named entity recognition system. In: Krish K (ed) Encyclopedia of library and information science. Marcel Decker, New York
8. Manning CD, Schutze H (1999) Foundations of statistical natural language processing. MIT Press, Cambridge, MA
9. Lothaire M (2005) Symbolic natural language processing. In: Applied combinatorics on words. Encyclopedia of mathematics and its applications (No. 105). Cambridge University Press, Cambridge, pp 164–209. Available from: Cambridge Books Online, http://dx.doi.org/10.1017/CBO9781107341005.004
10. Troussov A, Sogrin A, Judge J, Botvich D (2008) Mining socio-semantic networks using spreading activation technique. In: Proceedings of the I-KNOW '08 and I-MEDIA '08, Graz, 3–5 Sept 2008
11. Dale R, Moisl H, Somers HL (2000) Handbook of natural language processing. Marcel Dekker, New York
12. Korhonen A, Krymolowski Y (2006) A large subcategorisation lexicon for natural language processing applications. In: Proceedings of LREC

13. Stumme G (2002) Efficient data mining based on formal concept analysis. In: Proceedings of the DEXA '02 proceedings of the 13th international conference on database and expert systems applications, pp 534–546
14. Wilks Y, Stevenson M (1998) The grammar of sense: using part-of-speech tags as a first step in semantic disambiguation. Nat Lang Eng 4:135–143
15. Kuipers BJ (1984) Causal reasoning in medicine: analysis of a protocol. Cognit Sci 8:363–385
16. Danks D, Griffiths TL, Tenenbaum JB (2003) Dynamical causal learning. In: Becker S, Thrun S, Obermayer K (eds) Advances in neural information processing systems 15. MIT Press, Cambridge, pp 67–74
17. Trovati M, Bessis N (2015) An influence assessment method based on co-occurrence for topologically reduced big data sets. Soft Comput 1432–7643, pp 1–10

Chapter 8
Two Case Studies Based on Large Unstructured Sets

**Aaron Johnson, Paul Holmes, Lewis Craske, Marcello Trovati,
Nik Bessis, and Peter Larcombe**

Abstract In this chapter, we shall present two case studies based on large unstructured datasets. The former specifically considers the Patient Health Questionnaire (PHQ-9), which is the most common depression assessment tool, suggesting the severity and type of depression an individual may be suffering from. In particular, we shall assess a method which appears to enhance the current system in place for health professionals when diagnosing depression. This is based on a combination of a computational assessment method, with a mathematical ranking system defined from a large unstructured dataset consisting of abstracts available from PubMed. The latter refers to a probabilistic extraction method introduced in Trovati et al. (IEEE Trans ADD, 2015, submitted). We shall consider three different datasets introduced in Trovati et al. (IEEE Trans ADD, 2015, submitted; Extraction, identification and ranking of network structures from data sets. In: Proceedings of CISIS, Birmingham, pp 331–337, 2014) and Trovati (Int J Distrib Syst Technol, 2015, in press), whose results clearly indicate the reliability and efficiency of this type of approach when addressing large unstructured datasets. This is part of ongoing research aiming to provide a tool to extract, assess and visualise intelligence extracted from large unstructured datasets.

8.1 Introduction

In this chapter we discuss two case studies related to some large unstructured datasets aiming to evaluate and assess the capabilities of two approaches based on a combination of a variety of techniques and tools.

The first case study described in Sect. 8.2, which refers to one of the most common depression assessment methods, the Patient Health Questionnaire (PHQ-9) [4], regarded as a valuable resource for depression detection according to Cronbach's alpha [2]. Text mining techniques and a hierarchical system are integrated to provide

A. Johnson (✉) • P. Holmes • L. Craske • M. Trovati • N. Bessis • P. Larcombe
Department of Computing and Mathematics, University of Derby, Derby, UK
e-mail: aaronjohnjohnson@hotmail.com

© Springer International Publishing Switzerland 2015

111

M. Trovati et al. (eds.), *Big-Data Analytics and Cloud Computing*,
DOI 10.1007/978-3-319-25313-8_8

a suitable ranking and weighting of each item in the qPHQ-9 questionnaire. The second case study is discussed in Sect. 8.3, which focuses on the evaluation of method introduced in [16]. We will consider three different large datasets described in [13, 15, 16].

8.2 Case Study 1: Computational Objectivity in the PHQ-9 Depression Assessment

The Patient Health Questionnaire 9 (PHQ-9) consists of nine questions utilised by general practitioners and mental health professionals in the diagnosis of depression [4]. Each item of the questionnaire is part of the Diagnostic and Statistical Manual of Mental Disorders (DSM-IV) criteria [1]. As a consequence the PHQ-9 is widely regarded as an effective diagnostic method for depression [5]. More specifically, each item is scored from 0 ('not at all') to 3 ('every day') [7] with a general score of 27, as shown in Fig. 8.1 (Table 8.1).

8.2.1 Reliability and Validity Issues of the PHQ-9

The validity and reliability of the PHQ-9, and any other diagnostic method, needs to be constantly assessed.

In [2] an investigation on its validity for the diagnosis of depression in East Africa is discussed. This study was based on two stages, where first of all the PHQ-9 was used on patients, and subsequently, participants were also interviewed via the SCAN (Schedules for Clinical Assessment in Neuropsychiatry) questionnaire, designed by the World Health Organisation (WHO) [18]. More specifically, the SCAN questionnaire has been shown to be an accurate method to diagnose mental illnesses, which resulted in a valuable exercise in the criterion validity for the PHQ-9 [11].

Another measure of internal consistency is the Cronbach's alpha [12], which highlights the extent to which all items in a questionnaire measure the same concept. This further assesses the efficacy of PHQ-9 as a measure of depression [12]. It is computed as [17]

$$\alpha = \frac{K}{K-1} \left(1 - \frac{\sum_{i=1}^{K} \sigma_Y^2}{\sigma_X^2} \right) \tag{8.1}$$

where

- K is the number of items in the test, namely, 9 for PHQ-9 questionnaire,
- σ_X^2 is the variance of observed total questionnaire scores and

PATIENT HEALTH QUESTIONNAIRE-9 (PHQ-9)

Over the **last 2 weeks**, how often have you been bothered by any of the following problems? *(Use "✔" to indicate your answer)*	Not at all	Several days	More than half the days	Nearly every day
1. Little interest or pleasure in doing things	0	1	2	3
2. Feeling down, depressed, or hopeless	0	1	2	3
3. Trouble falling or staying asleep, or sleeping too much	0	1	2	3
4. Feeling tired or having little energy	0	1	2	3
5. Poor appetite or overeating	0	1	2	3
6. Feeling bad about yourself — or that you are a failure or have let yourself or your family down	0	1	2	3
7. Trouble concentrating on things, such as reading the newspaper or watching television	0	1	2	3
8. Moving or speaking so slowly that other people could have noticed? Or the opposite — being so fidgety or restless that you have been moving around a lot more than usual	0	1	2	3
9. Thoughts that you would be better off dead or of hurting yourself in some way	0	1	2	3

FOR OFFICE CODING ___0___ + _____ + _____ + _____

=Total Score: _____

If you checked off **any** problems, how **difficult** have these problems made it for you to do your work, take care of things at home, or get along with other people?

Not difficult at all □	Somewhat difficult □	Very difficult □	Extremely difficult □

Developed by Drs. Robert L. Spitzer, Janet B.W. Williams, Kurt Kroenke and colleagues, with an educational grant from Pfizer Inc. No permission required to reproduce, translate, display or distribute.

Fig. 8.1 The PHQ-9 Questionnaire

- σ_Y^2 is the variance of component i for the current sample of people [17].

Moreover, Cronbach's alpha is contained in the interval [0, 1], whose categorisation is shown in Table 8.2.

Table 8.1 Scoring categories for PHQ-9 results

Outlined depression severity	PHQ-9 score
1–4	None
5–9	Mild
10–14	Moderate
15–19	Moderately severe
20–27	Severe

Table 8.2 Characterisation of the Cronbach's alpha

$\alpha \geq 0.9$	Excellent
$0.7 \geq \alpha < 0.9$	Good
$0.6 \geq \alpha < 0.7$	Acceptable
$0.5 \geq \alpha < 0.6$	Poor
$\alpha < 0.5$	Unacceptable

Cronbach's alpha has been assessed as 0.85, suggesting that the PHQ-9 is a good resource for detecting depression [2]. However, it is clear that depression detection can still be enhanced in order to upgrade it to the 'excellent' category, that is when $\alpha \geq 0.9$.

Furthermore, reliability estimates can be calculated, which provides a useful insight into the interpretation on the associated reliability [12], where a positive correlation suggests high reliability and no correlation no reliability at all. In particular, let $r \in [0, 1]$ be the reliability estimate and IME be the index of measurement error. We then have that

$$IME = 1 - r^2 \tag{8.2}$$

Note that, as the reliability estimate increases, the fraction of a test score related to error will decrease.

8.2.2 Analytic Hierarchy Process: Defining a Weighting System

Analytic hierarchy process (AHP) is a technique aiming to provide a decision-making process with structured intelligence [3]. AHP has been extensively analysed with a wide set of applications to a variety of multidisciplinary contexts. The motivation is based on the observation that a decision process should independently consider each factor contributing to the overall objective. AHP is utilised to assess the importance of specific contributors, which are likely to exert an influence on an objective, by carrying out a pairwise comparison between each and every contributing elements. This process utilises a rating scale, as depicted in Table 8.3, to determine the importance of each element when compared with the other ones [3].

Table 8.3 The fundamental scale for pairwise comparisons

Intensity of importance	Definition	Explanation
1	Equal importance	Two elements contribute equally to objective
3	Moderate importance	Experience and judgement slightly favour one element over the other
5	Strong importance	Experience and judgement strongly favour one element over the other
7	Very strong importance	Experience and judgement very strongly favour one element over the other
9	Extreme importance	The evidence favouring one element over another is of the highest possible order of affirmation

This scale has been validated for effectiveness, not only in many applications but also through theoretical comparisons with a large number of other scales [10].

8.2.2.1 PHQ-9 Analysis via the Analytic Hierarchy Process

AHP can be applied to the PHQ-9 questionnaire by realising that the desired objective is the relevant diagnosis, and the contributors towards this objective are the nine questions.

To achieve objectivity, selected experts in the field would consider each pairwise combination of questions so that they could decide and rate which of the questions they consider more important. Once all the questionnaires have been completed, an overall mean average can be computed, allowing the construction of a final analysis pairwise comparison table.

8.2.2.2 Advantages of AHP

AHP is an efficient method, which can be interactively modified at any stage of development. The method allows a thorough investigation of the problem, so that any discrepancies can be addressed effectively [6]. Furthermore, the AHP questionnaire could be distributed to a large number of experts representing particular statistical cross sections of a population.

8.2.3 A Text Mining Approach

In order to optimise above method more efficiently, text mining techniques were implemented to automate the use of the AHP described in Sect. 8.2.2.1.

Table 8.4 A selection of keywords within each question in PHQ-9

Question	Keywords
1	Pleasure
	Interest
2	Hopeless
	Feeling down
	Depressed
3	Asleep
	Sleep
4	Tired
	Energy
5	Appetite
	Overeating
6	Failure
	Feeling bad
7	Concentrating
8	Moving slowly
	Fidgety
	Restless
9	Death
	Hurting

More specifically, each question in the PHQ-9 questionnaire contains specific keywords; Table 8.4 provides a selection.

The main assumption is that such collections of keywords give an indication of the type and frequency of their usage. More specifically, the more often they appear, the more likely they are to be relevant in depression detection via the PHQ-9.

PubMed [14] contains over 24 million citations and abstracts for biomedical literature, life science journals and on-line books.

Approximately 250,000 abstracts from PubMed were identified by considering the following keywords: 'PHQ-9' and 'depression'. Table 8.5 depicts the trend of their co-occurrences in these abstracts over a 15 year period.

Subsequently, the combination of the keywords corresponding to each of the nine questions was extracted from the above abstract. Table 8.6 shows their number of co-occurrence.

Subsequently, each of the items in Table 8.6 was taken in pairs, which was then fed into the AHP method to assess the importance of the questions.

Let a and b be the number of occurrences of the keywords associated with any two questions, and let $a > b$. We then consider Algorithm 1.

Subsequently, by following the AHP steps, we obtained that the order of the questions is shown in Table 8.7

The above results were assessed by a number of health professionals, who have agreed on the suggested ordering. Furthermore, the weight of each question

Table 8.5 Scoring categories for PHQ-9 results

Year	PHQ-9/Depression keywords count
2015	121
2014	232
2013	174
2012	144
2011	92
2010	71
2009	65
2008	41
2007	32
2006	14
2005	15
2004	10
2003	3
2002	1
2001	1

Table 8.6 Number of instances of keywords associated with the questions in the PHQ-9 questionnaire

Question 1	404
Question 2	1592
Question 3	2210
Question 4	231
Question 5	101
Question 6	371
Question 7	49
Question 8	40
Question 9	514

Algorithm 1 The fundamental scale for pairwise comparisons

1: **if** $a \approx b$ **then**
2:     output = 1
3: **end if**
4: **if** $4a/5 \leq a - b \leq 1$ **then**
5:     output = 3
6: **end if**
7: **if** $3a/5 \leq a - b < 4a/5$ **then**
8:     output = 5
9: **end if**
10: **if** $2a/5 \leq a - b < 3a/5$ **then**
11:     output = 7
12: **end if**
13: **if** $a/5 \leq a - b < 2a/5$ **then**
14:     output = 9
15: **end if**
16: **return** X

Table 8.7 The ranking of the questions of the PHQ-9 questionnaire

Question number
3
2
9
1
6
4
5
7
8

Table 8.8 The weighting of the questions of the PHQ-9 questionnaire

Question number	Percentage
1	7
2	28
3	40
4	4
5	2
6	5
7	3
8	2
9	9

was determined by renormalising the number of occurrences of the corresponding keywords, as depicted in Table 8.8, which proved more difficult to agree upon by the health professionals, with average of 40 %. This can be easily explained by the fact that this step would require a much deeper understanding of the issues regarding depression assessment compared to their ranking.

8.3 Case Study 2: Evaluation of Probabilistic Information Extraction from Large Unstructured Datasets

The identification and extraction of probabilistic information is of crucial importance in the assessment and interpretation of Big Data. In this section, the method introduced in [16] is evaluated to assess its extraction power of probabilistic relations between concepts, based on the datasets introduced in [13, 15, 16]. We will show that this further supports the preliminary evaluation carried out in [16], as well as its potential to successfully address specific challenges posed by unstructured and highly dynamic data sources.

8.3.1 Description of the Method

A text and data mining approach is introduced in [16] to firstly extract the concepts and their mutual relationships, which are subsequently used to populate (fragments of) a network, whose topological and dynamical properties are investigated.

8.3.1.1 Description of Text and Data Patterns

The text mining approach is based on the following extraction types:

- **Text patterns**. These identify specific text fragments that indicate the existence of probabilistic information between concepts.
- **Data patterns**. On the other hand, these refer to any data that can be extracted and identified, which show well-defined properties, not necessarily captured by text patterns.

Furthermore, in order to be identified by text patterns, concepts and their mutual relations must appear within the same text fragments or be linked by specific database relationships. The output is a set of quadruples (NP1, MOD, keyword, NP2) where:

- NP1 and NP2 are the *noun phrases*, i.e. phrases with a noun as the head word, containing one or more concepts suitably defined [16].
- keyword refers to any collection of probabilistic terms contained in the ontology discussed in [16].
- MOD is the keyword *modality*, which specifies whether the corresponding sentence refers to either a relation or a 'non-relation' in probabilistic terms.

For example, the sentence

A small increase in the atmospheric humidity level is **not** necessarily linked with an increased risk of thunderstorm.

suggests a lack of any probabilistic relationship between *atmospheric humidity level* and *increased risk of thunderstorm*, or in other words, we have an independence relation between the two concepts.

This type of independence relationship is directly specified as follows:

- positive MOD + independence keyword = independence (probabilistic) relation
- positive MOD + dependence keyword = dependence (probabilistic) relation
- negative MOD + independence keyword = dependence (probabilistic) relation
- negative MOD + dependence keyword = independence (probabilistic) relation

Figures 8.2 and 8.3 depict the general architecture of the above steps.

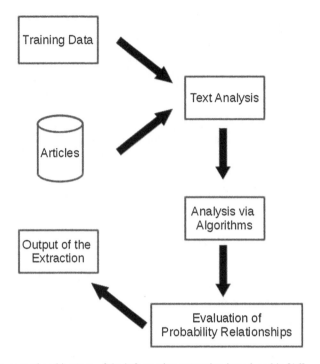

Fig. 8.2 The general architecture of the information extraction introduced in [16]

8.3.2 Network Extraction Method

Different concepts and mutual relationships define a relational network $G(V, E)$, where V is the *node-set*, and E is the *edge-set* [14]. Such networks have been shown to exhibit a scale-free structure characterised by the existence of several giant clusters.

The steps implemented in the assessment and evaluation of the probability level between two concepts include a variety of aspects, including the general concepts that are shared among the nodes, and the nature of the linking relations, in terms of known probabilistic relationships. Furthermore, temporal information can also give an insight into the type of information aggregation required [16].

8.3.3 Description of Datasets

The datasets utilised in the evaluation described in Sect. 8.3.4 are fully described in [13, 15, 16], which include:

• 150 biomedical articles and 300 abstracts, freely available from PubMed [8];

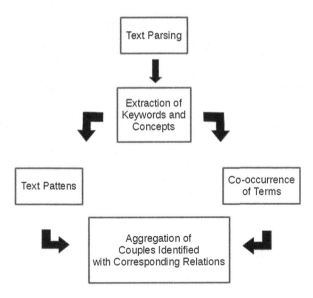

Fig. 8.3 The specific steps of the text analysis discussed in Sect. 8.3.1.1

- Earthquake instances collected by the European-Mediterranean Seismological Centre;
- Air accidents and near misses.

In particular, Figs. 8.4 and 8.5 depict fragments of the networks generated by the above datasets.

8.3.4 Evaluation

As specified in [16], the context parameters w_i for $i = 1, 2, 3$ were defined by considering the following concepts:

- `cancer` and `cancer_of_<TYPE>`,
- `infection` and `infection_of_<TYPE>`,
- `earthquake` and `earthquake_of_<TYPE>`,
- `air_accidents` and `air_accidents_of_<TYPE>`,

where `<TYPE>` is a specific type related to the corresponding concepts.

In particular, the assumption that the joint probability of any concept and its type is greater than or equal to 0.9 [16] produced that $w_1 = 0.51$, $w_2 = 0.27$ and $w_3 = 0.59$.

The well-established pharmacogenic associations consisting of approximately 664 associated couples were investigated[9], as shown in Table 8.9, whose joint probability between any of the couples was assumed to be above 0.8 [16].

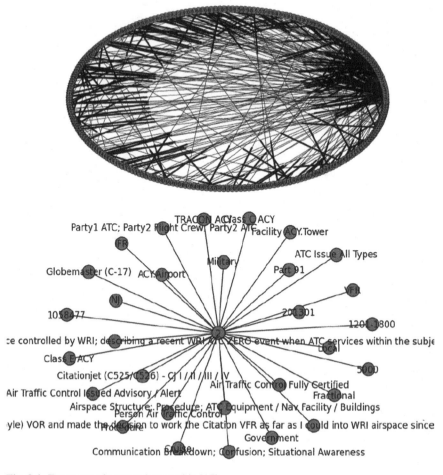

Fig. 8.4 Fragments of a network created in [15]

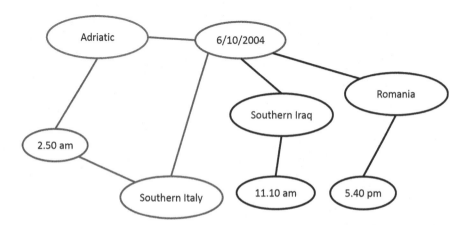

Fig. 8.5 Fragment of the relational network generated in [13]

Table 8.9 A small selection of well-known pharmacogenic associations as in [9]

Question number	Percentage
Abacavir	HLA-B
Abiraterone	CYP17A1
Abiraterone	CYP1A2
Acenocoumarol	CYP2C9
Acenocoumarol	VKORC1
Acetaminophen	CYP1A2
Acetaminophen	CYP2D6
Afatinib	ABCB1
Afatinib	ABCG2
Afatinib	CYP1A2

Table 8.10 Evaluation of the dataset described in [16]

Concept A	Concept B	P(concept A, concept B)	Human evaluation
Cancer	Smoking	0.69	0.8–0.9
Heart disease	Beta-blockers	0.47	0.4–0.6
Organophosphate	pyrethroid	0.18	0.2–0.5
Infection	Ocular toxoplasmosis	0.31	0.4–0.5

Subsequently, we considered approximately 400 abstracts regarding topics on pharmacogenic associations, and 63 % of them were identified with a joint probability above the threshold.

The following concepts were utilised to assess the dataset described in [16]:

- *cancer* and *smoking*
- *heart disease* and *beta-blocker,*
- *organophosphate* and *pyrethroid,*
- *infection* and *ocular toxoplasmosis.*

These were used to evaluate the above method, as depicted in Table 8.10.

The dataset described in [13] was subsequently investigated, by isolating the following concepts:

- *Adriatic (area)* and *Southern Italy,*
- *Aegean Sea* and *Central Italy,*
- *Estonia* and *Crete.*

The results are comparable with the human evaluation described in [13], showing accuracy, as shown in Table 8.11.

Finally, from the dataset described in [15], we looked at the following concepts:

- *destination unclear* and *unexpected heading,*
- *geographical coordinates* and *out waypoint,*
- *multiple TCAS RA events* and *go arounds.*

Table 8.11 Evaluation of the dataset introduced in [13]

Concept A	Concept B	P(concept A, concept B)	Human evaluation
Adriatic (area)	Southern Italy	0.73	0.80–0.90
Aegean Sea	Central Italy	0.41	0.20–0.30
Estonia	Crete	0.09	0.05–0.10

Table 8.12 Evaluation of the dataset introduced in [15]

Concept A	Concept B	P(concept A, concept B)	Human evaluation
Destination unclear	Unexpected heading	0.63	0.60–0.90
Geographical coordinates	Out waypoint	0.41	0.50–0.60
Multiple TCAS RA events	Go arounds	0.83	0.70–0.80

Table 8.12 shows the evaluation results, showing again promising accuracy.

8.4 Conclusion

In this chapter, we have discussed two study cases related to Big Data. The former based on the PHQ-9 questionnaire, and the latter on the method introduced in [16]. Both approaches are part of a wider line of inquiry to provide an effective, accurate and scalable set of tools to obtain intelligence from Big Data. The evaluations discussed in this chapter clearly show potential for a variety of far-reaching applications, which could have important implications in multiple research fields.

Acknowledgements This research was partially supported by the University of Derby Undergraduate Research Scholarship Scheme.

References

1. American Psychiatric Association (1994) Diagnostic and statistical manual of mental disorders, 4th edn. American Psychiatric Association, Washington, DC
2. Gelaye B et al (2013) Validity of the patient health questionnaire-9 for depression screening and diagnosis in East Africa. Psychiatry Res Dec 210(2). doi:10.1016/j.psychres.2013.07.015
3. Harker PT (1986) Incomplete pairwise comparisons in the analytic hierarchy process. University of Pennsylvania, Philadelphia
4. Kroenke K, Spitzer R (2002) The PHQ-9: a new depression diagnostic and severity measure. Psychiatric Ann 32:1–6
5. Manea L, Gilbody S, McMillan D (2012) Optimal cut off score for diagnosing depression with the Patient Health Questionnaire (PHQ-9): a meta-analysis. CMAJ 184(3):E191–E196
6. Murphy RO Prof (2013) Decision theory: rationality, risk and human decision making

7. Thomas H (2013) Patient Health Questionnaire (PHQ-9). Patient, Available from http://patient.info/doctor/patient-health-questionnaire-phq-9. Last accessed 17 Jun 2015

8. Pubmed, Available from http://www.ncbi.nlm.nih.gov/pubmed. [15 April 2014]

9. PharmGKB, Available from https://www.pharmgkb.org/. [15 April 2014]

10. Saaty TL et al (1990) How to make a decision: the analytic hierarchy process, Pittsburgh

11. Salkind N, Rasmussen K (2007) Encyclopedia of measurement and statistics. SAGE, Thousand Oaks. doi:10.4135/9781412952644

12. Tavakol M, Dennick R (2011) Making sense of Cronbach's Alpha. Int J Med Educ 2:53–55

13. Trovati M (2015) Reduced topologically real-world networks: a big-data approach. Int J Distrib Syst Technol (IJDST) 6(2):13–27

14. Trovati M, Bessis N (2015) An influence assessment method based on co-occurrence for topologically reduced big datasets. Soft computing. Springer, Berlin/Heidelberg

15. Trovati M, Bessis N, Huber A, Zelenkauskaite A, Asimakopoulou E (2014) Extraction, identification and ranking of network structures from data sets. In: Proceedings of CISIS, Birmingham, pp 331–337

16. Trovati M, Bessis N, Palmieri F, Hill R (under review) Extracting probabilistic information from unstructured large scale datasets. IEEE Syst J

17. Vale L, Silcock J, Rawles J (1997) An economic evaluation of thrombolysis in a remote rural community. BMJ 314:570–572

18. Wing J et al (1990) SCAN schedules for clinical assessment in neuropsychiatry. Arch Gen Psychiatry 47(6):589–593. doi:10.1001/archpsyc.1990.01810180089012

Chapter 9
Information Extraction from Unstructured Data Sets: An Application to Cardiac Arrhythmia Detection

Omar Behadada

Abstract In this chapter, we will discuss a case study, which semi-automatically defines fuzzy partition rules to provide a powerful and accurate insight into cardiac arrhythmia. In particular, this is based on large unstructured data sets in the form of scientific papers focusing on cardiology. The information extracted is subsequently combined with expert knowledge, as well as experimental data, to provide a robust, scalable and accurate system. The evaluation clearly shows a high accuracy rate, namely, 92.6 %, as well as transparency of the system, which is a remarkable improvement with respect to the current research in the field.

9.1 Introduction

Cardiovascular diseases are one of the most worrying health issues and the largest cause of mortality in the world, based on the World Health Report 2013 [1]. Therefore, a low-cost and high-quality cardiac assessment would be a very valuable contribution to the well-being of every individual in our societies.

In particular, being able to detect cardiac arrhythmia is a very promising area, since *premature ventricular contraction* (PVC) is an effective predictor of sudden death. Several studies over the past decade have focused on methods and algorithms for detection and significance of cardiac arrhythmias, aiming to achieve a good classification rate [2].

More specifically, these include Bayesian classifiers, decision trees, neural and rule-based learners [2]. However, classification approaches with good classification rates usually have a low degree of interpretability preventing the user (such as a cardiologist) from fully taking advantage of such methods.

O. Behadada (✉)
Department of Biomedical Engineering, Faculty of Technology, Biomedical Engineering Laboratory, University of Tlemcen, Tlemcen, Algeria
e-mail: omar.behadada@gmail.com; o_behadada@mail.univ-tlemcen.dz

© Springer International Publishing Switzerland 2015   127
M. Trovati et al. (eds.), *Big-Data Analytics and Cloud Computing*,
DOI 10.1007/978-3-319-25313-8_9

The *interpretability*, also referred to as *comprehensibility*, *intelligibility* or *transparency*, of knowledge-based systems is of crucial importance [3]. Such systems consist of two main elements, a reasoning mechanism and a knowledge base (KB) [4].

In fact, in order to maintain, manipulate and update the knowledge captured by such systems, their associated KBs should be as interpretable as possible, so that users are confidently facilitated in the assessment and decision-making process. However, expert knowledge acquisition remains a hard and critical task, and it is considered as a bottleneck in the expert system modelling process. Furthermore, some parts of expert knowledge remain at an unconscious level and are difficult to formalise [5]. However, systems can also be built using other sources of information such as experimental data, which are likely to give an accurate overview of the different parameters. There are, in the specialised literature, a wide range of algorithms and machine-learning (ML) techniques for model identification that are driven by the improvement of accuracy indices [6]. Some of these algorithms can be used as knowledge induction techniques [7, 8].

Another valuable source of knowledge can be obtained from articles and texts published in scientific journals, which include the most critical knowledge, where many experts share their results, analyses and evaluations. However, since such textual information is typically very large, scientists are faced with great amount of information which poses a huge computational and implementation challenge. In modern medicine, huge amounts of data are continuously generated, but there is a widening gap between data acquisition and data comprehension. It is often impossible to process all of the data available and to make a rational decision on basic trends. Therefore, there is an increasing need for intelligent data analysis to facilitate the creation of knowledge to support clinicians in making decision.

In this chapter, we discuss a case study based on a semi-automatic method to identify fuzzy partition rules applied to cardiac arrhythmia detection. This combines an automated information extraction from textual sources with expert elicitation to create a robust, scalable and accurate knowledge-based system which provides a crucial insight into arrhythmia detections.

The chapter is structured as follows: in the rest of this section, we discuss the relevant medical background; in Sect. 9.3 we introduce the text mining method to extract information for the generation of fuzzy partitions, which are subsequently analysed and evaluated in Sects. 4, 5, 6 and 7. Finally in Sect. 8, we discuss the main findings and future research directions.

9.2 Background

Electrocardiogram (ECG) reflects activity of the central blood circulatory system. An ECG (Fig. 9.1) signal can provide a great deal of information on the normal and pathological physiology of heart activity. Thus, ECG is an important non-invasive clinical tool for the diagnosis of heart diseases [9, 10]. Early and quick detection

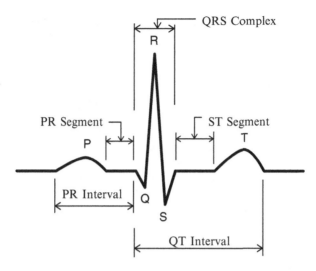

Fig. 9.1 Standard ECG beat

and classification of ECG arrhythmia is important, especially for the treatment of patients in the intensive care unit [1]. For more than four decades, computer-aided diagnostic (CAD) systems have been used in the classification of the ECG resulting in a huge variety of techniques [11]. Included in these techniques are multivariate statistics, decision trees, fuzzy logic [12], expert systems and hybrid approaches [8]. In designing of CAD system, the most important step is the integration of suitable feature extractor and pattern classifier such that they can operate in coordination to make an effective and efficient CAD system [13].

Much of the information available in the real world is qualitative rather than quantitative, and as a consequence, it cannot be precisely measured. In fact, human communication is inherently vague, imprecise and uncertain. Fuzzy logic was introduced to mathematically model this ambiguity and vagueness [14] and explains the varied nature of ambiguity and uncertainty that exist in the real world [8]. As opposed to classical logic and set theory, where information is more often expressed in quantifying propositions, fuzzy logic can address the concept of partial truth or, in other words, truth values between being completely true and completely false.

Fuzzy set theory [15, 16] has several applications in data mining [17], clustering [18], association rules [19] and image retrieval [20].

9.3 Automated Extraction of Fuzzy Partition Rules from Text

Text mining [21] is a branch of computer science whose objectives include the extraction, identification and analysis of relevant and accurate information from textual sources. Even though there has been steady and successful progress in addressing the above challenges, text mining research is still very much expanding to provide further state-of-the-art tools to enhance accuracy, scalability and flexibility.

The extraction of information from text is typically a complex task due to the highly ambiguous nature of human language. In particular, based on the general context and the given semantic information, a variety of text mining techniques can be utilised [21].

In this chapter, we will discuss a grammar-based text extraction approach, or *text patterns*, which depends on a set of rules identifying sentences with a determined structure. More specifically, we consider text patterns of the form 'NP, verb, NP', where NP refers to the noun phrases and verb to the linking verb [22]. For example, sentences such as 'the occurrence of PVCs has been shown to be associated with electrolytes' are identified to extract a relationship between *PVCs* and *electrolytes*. The effectiveness of this approach is particularly clear when syntactic properties of a sentence are investigated, by using suitable parsing technology [23]. In particular, the syntactic roles of the different phrasal components are critical in extracting the relevant information, and they can contribute towards a full understanding of the type of relationship. Furthermore, we also apply basic sentiment analysis techniques to identify the mood embedded within text fragments defined by specific keywords. Table 9.1 shows a small selection of such keywords used in our approach.

All the journals specialising on cardiology were downloaded from PubMed [24], as XML files, which were analysed as follows:

- We identified articles from the above journals containing the keywords:

    - Premature ventricular contractions or PVCs
    - Premature ventricular complexes
    - Ventricular premature beats
    - Extrasystoles

- The identified articles were both lexically and syntactically analysed via the Stanford Parser [23].

Table 9.1 A selection of keywords used

Negative keywords	Positive keywords	Uncertain keywords
Bad	Satisfactory	Unpredictable
Negative	Enhancing	Possible
Underestimate	Advantage	Somewhat
Unsafe	Benefit	Precautions
Unwelcome	Good	Speculative
Tragic	Excellent	Confusing
Problematic	Great	Fluctuation

Table 9.2 Example of relation extraction

Keywords in relation extraction	Sentences identified
'PVC', 'PVCs', 'imbalances'	'PVCs can be related to electrolyte or other metabolic imbalances'
'PVC', 'death'	'70 mmol/L and T2DM significantly increase risk of PVC and sudden cardiac death; the association between sMg and PVC may be modified by diabetic status'
'Premature ventricular complexes', 'PVC', 'PVCs', 'atrial', 'ventricular', 'beat', 'missed', 'premature'	'The system recognises ventricular escape beats, premature ventricular complexes (PVC), premature supraventricular complexes, pauses of 1 or 2 missed beats, ventricular bigeminy, ventricular couplets (2 PVCs), ventricular runs (> 2 PVCs), ventricular tachycardia, atrial fibrillation/flutter, ventricular fibrillation and asystole'

- Subsequently, a grammar-based extraction identifies the relevant information based on the above keywords as well as on sentiment analysis [25]. More specifically, only sentences with one or more of the above keywords, including those in Table 9.1, in the NPs were extracted.

9.3.1 Text Mining Extraction Results

The output of the extraction included the set of keywords found in each text fragment (usually a sentence) which was also extracted; see Table 9.2 for an example. An assessment of this type of information extraction from text was carried out, and the automatic extraction was then compared with a manual one which produced a recall of 68 % and a precision of 75 %.

9.4 Data Preparation

Following the text analysis described above, data obtained from MIT-BIH [26] was considered. In particular this refers to the patients as shown in Table 9.3. The R peaks of the ECG signals were detected using the Tompkins algorithm [27], an online real-time QRS detection algorithm. This algorithm reliably detects QRS complex using slop, amplitude and width information. This algorithm automatically adjusts thresholds and parameters periodically to the standard 24 h MIT-BIH arrhythmia database; this algorithm correctly detects 99.3 % of QRS complex.

From patients with cardiac arrhythmia, taken from MIT-BIH database, only patients with premature ventricular contraction (PVC) beats, premature arterial

Table 9.3 Evaluation data taken from the MIT-BIH database

Record				
	N	A	J	V
101	1860	3	–	–
103	2082	2	–	–
104	163	–	–	2
105	2526	–	–	41
106	1507	–	–	520
107	–	–	–	59
108	1739	4	–	17
109	–	–	–	38
111	–	–	–	1
112	2537	2	–	–
113	1789	–	–	–
114	1820	10	2	43
115	1953	–	–	–
116	2302	1	–	109
117	1534	1	–	–
118	–	96	–	16
119	1543	–	–	444
121	1861	1	–	1
122	2476	–	–	–
123	1515	–	–	3
124	–	2	29	47
200	1743	30	–	826
201	1625	30	1	198
202	2061	36	–	19
203	2529	–	–	444
205	2571	3	–	71
207	–	107	–	105
208	1586	–	–	992
209	2621	383	–	1

contraction (PAC) beats and premature junctional contraction (PJC) beats were selected. In fact, they provide the best quality of records, and more specifically PVC is a predictive element of the CA sudden death (Fig. 9.2) [11, 28].

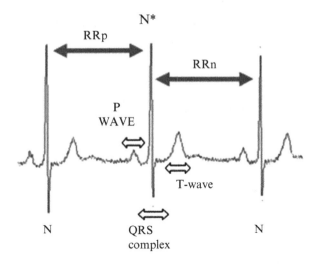

Fig. 9.2 Standard ECG beat

Data set:

Class	Normal	PVC	PAC	PJC
Number of samples	60,190	6709	2130	83

9.4.1   Feature Selection

The feature vector x, which is used for recognition of beats, was identified according to the RR interval of the beat RRp (i.e. the difference between the QRS peak of the present and previous beat), the ratio $r = RR1\text{-to-}RR0(RRn$ is calculated as the difference between the QRS peak of the present and following beat, as depicted in Fig. 9.3) and finally the QRS width w (determined by the Tompkins algorithm [27]). As a consequence, each beat is stored as 3-element vector. Table 9.4 provides the most relevant parameters used in this type of analysis.

9.5   Fuzzy Partition Design

In this section, the steps related to membership function extraction from both experts' input and experimental data are discussed. The initial step considers the extraction from the former, and subsequently some approaches for membership function design from data are introduced [16, 29].

FUZZY PARTITIONS|RULES

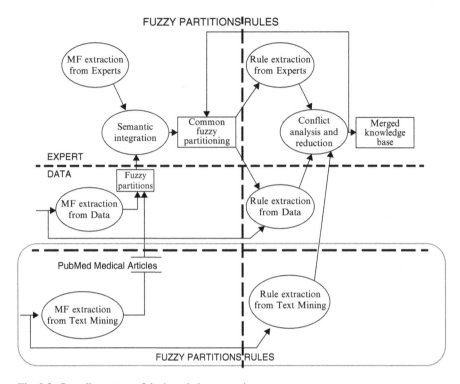

Fig. 9.3 Overall structure of the knowledge extraction process

Table 9.4 The various descriptors

Attributes	Meaning
RR precedent: RR0	The distance between the peak of the beat R and peak of the precedent beat R
RR next : RRn	The between the peak of the present R beat and the peak of the following R beat
QRS *complex*	Beginning of the Q wave and the end of the S wave
Comp	The ratio RR0/RRs
PP	Pic to pic of the R wave of the QRS complex
Energy	Energy of the QRS complex

In particular, Fig. 9.3 depicts the overall structure of the extraction process from expert knowledge, data and textual information. Furthermore, fuzzy logic enables the merging and management of these three types of knowledge, where the fuzzy partition design aims to define the most influential variables, according to the above knowledge.

The core of this process is the rule base definition and integration, during which the expert is invited to provide a description of the system behaviour, by providing specific system knowledge as linguistic rules (*expert rules*). In addition, rules are induced from data (*induced rules*) according to the properties of fuzzy partition. As a consequence, rule comparison can be carried out at a linguistic level, and subsequently, both types of rules are integrated into a unique knowledge base.

The main assumption of this approach is that expert knowledge is assumed to be provided, at least, at a minimal level, regarding the linguistic variables of the system. In particular, for a given variable, a specific domain of interest should be defined, so that decisions are facilitated by a given, and small, number of linguistic terms. On the other hand, it is also possible to consider parameterised functions in expert knowledge extraction, where the main task is the adjustment of the parameters by means of statistical methods. In our case, as discussed above, minimum information on membership function definition will be assumed such as the definition of universes, number of terms and, sometimes, prototypes of linguistic labels (modal points). This approach is a reduced version of the interval estimation method, reducing the interval to a single point.

The knowledge base is divided into two main parts, the *data base* (DB) and the *rule base* (RB). The former usually includes the description of the linguistic variables (number, range, granularity, membership functions) and the scaling or normalisation function related to those variables. Regarding the generation from data of the fuzzy partitions, most of the previous contributions in the field consider the automatic design of the data base as one of the steps in the definition of the overall knowledge base of a fuzzy rule-based system. This situation is slightly modified in this approach, since rules will be generated on the basis of fuzzy partitions that integrate both expert and data-based knowledge.

The automatic generation of fuzzy partitions is based on the definition of the most appropriate shapes for the membership functions, determining the optimum number of linguistic terms in the fuzzy partitions (i.e. the granularity) and/or locating the fuzzy sets into the universe of the variable.

In terms of the DB design, only some selected approaches are applicable in this context.

In particular, the preliminary design [30] involves extracting the DB a priori by induction from the available data set. This process is usually performed by non-supervised clustering techniques, by data equalisation techniques [31] or by an ascending method based on fuzzy set merging. These approaches consider only information related to input variables, and no assumption about the output is made. Furthermore, embedded design derives the DB using an embedded basic learning method [2].

A similar method is simultaneous design [30] which is not suitable in this context as it usually produces a much more complex process where the computational effort is partly useless since only the fuzzy partitions are considered. A posteriori design could also be considered in further steps of the process, except during the definition of the common fuzzy partitioning. In addition, our interpretability requirements need some specific properties in the definitions of the partitions. More specifically, techniques generating multidimensional clusters cannot be applied, since we need one-dimensional membership functions, obtained by independently partitioning the universe of each variable. As a general approach, it is possible to use any one-dimensional optimisation technique if it includes some semantic constraints.

In the case of embedded design, when search techniques are used to define the DB, it is possible to ensure the integrity properties, so that they include interpretability measures in the objective function, thus guiding the trek to good solutions. These usually include measures of completeness, consistency, compactness or similarity.

9.5.1 Criteria for the Evaluation of Fuzzy Partitions

The evaluation of fuzzy partitions involves two different aspects: linguistic properties and partitioning properties. Linguistic properties are solely related to the shape of the fuzzy sets and the relations between fuzzy sets defined for the same variable. Their assessment does not involve data. On the other hand, the partitioning properties refer to the level of matching of the partition with the data from which the partitions have been derived. It is only applicable for partitions derived from data.

The main linguistic properties, which have been considered at the beginning of this section, are guaranteed by the strong fuzzy partition constraints. In this section, partitioning properties are discussed, which include the definition of fuzzy sets and the distribution of data. In order to achieve this, several criteria related to clustering techniques can be applied.

The first remark is that only the methods relying on the distribution of data itself are considered, excluding those based on an input–output relation. Consequently, we focus on those methods that are usually applied in unsupervised clustering and not in supervised clustering.

As the criteria are defined to evaluate the matching between a data distribution and a partition, we will consider the assignment of elements of the data set to each of the elements of the partition (assuming a fuzzy partition). Consequently, we will represent the degree of membership of the k-th element of the data set to the i-th element of the fuzzy partition as u_{ik}. The number of elements of the fuzzy partition will be represented by c, while the cardinality of the data set will be n. With this definition, the partition coefficient is obtained as

$$PC = \frac{\sum_{k=1}^{n} \sum_{i=1}^{c} u_{ik}^2}{n}$$

and the partition entropy as

$$PE = -\frac{1}{n} \left\{ \sum_{k=1}^{n} \sum_{i=1}^{c} [u_{ik} \log (u_{ik})] \right\}$$

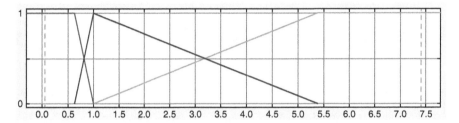

Fig. 9.3 Fuzzy partition RR0 from k-means algorithm

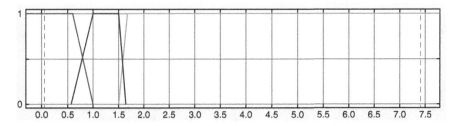

Fig. 9.4 Fuzzy partition RRs from expert and TM

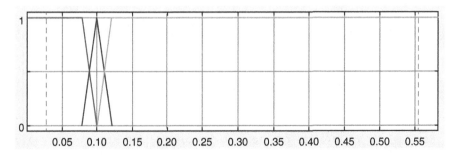

Fig. 9.5 Fuzzy partition QRS from expert and TM

Recently, a different index has been proposed by Chen [32]:

$$Ch = \frac{1}{n}\sum_{k=1}^{n}\max_{i} u_{ik} - \frac{2}{c(c-1)}\sum_{i=1}^{c-1}\sum_{j=i+1}^{c}\frac{1}{n}\sum_{k=1}^{n}\min\left(u_{ik}, u_{jk}\right)$$

The three indices described above can be applied to any partition, independently, of the derivation method. According to these criteria, a good partition should minimise the entropy and maximise the coefficient partition and the Chen index (Figs. 9.4, 9.5, 9.6, 9.7, 9.8, 9.9 and Tables 9.5, 9.6, 9.7, 9.8, 9.9, 9.10).

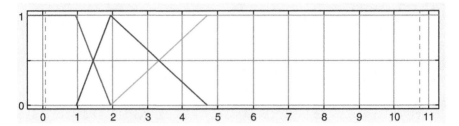

Fig. 9.6 Fuzzy partition COMP from k-means algorithm

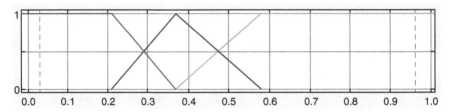

Fig. 9.7 Fuzzy partition PP from k-means algorithm (10^3)

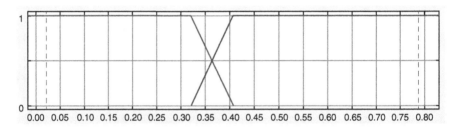

Fig. 9.8 Fuzzy partition energy from HFP algorithm

Table 9.5 *RR0*, fuzzy partition quality (3 labels)

Partition	Partition coefficient(max)	Partition entropy(min)	Chen index(max)
HFP	0.77513	0.33406	0.77094
Regular	0.69936	0.47483	0.74665
K-means	0.79948	0.30842	0.80794
Expert and TM	0.78262	0.32504	0.78647

Table 9.6 RRs fuzzy partition quality (3 labels)

Partition	Partition coefficient(max)	Partition entropy(min)	Chen index
HFP	0.77378	0.33614	0.76980
Regular	0.69705	0.47751	0.74360
K-means	0.77121	0.34723	0.77941
Expert and TM	0.78300	0.32441	0.78668

Table 9.7 QRS fuzzy partition quality (3 labels)

Partition	Partition coefficient(max)	Partition entropy(min)	Chen index(max)
HFP	0.76812	0.34495	0.76312
Regular	0.66966	0.50649	0.70024
K-means	0.82214	0.26975	0.82707
Expert and TM	0.84540	0.22046	0.82512

Table 9.8 COMP fuzzy partition quality (3 labels)

Partition	Partition coefficient(max)	Partition entropy(min)	Chen index(max)
HFP	0.68847	0.45294	0.67092
Regular	0.71224	0.45990	0.75945
K-means	0.89473	0.18398	0.90993
Expert and TM	0.81362	0.29566	0.82971

Table 9.9 PP fuzzy partition quality (3 labels)

Partition	Partition coefficient(max)	Partition entropy(min)	Chen index(max)
HFP	0.68847	0.45294	0.67092
Regular	0.71224	0.45990	0.75945
K-means	0.89473	0.18398	0.90993
Expert and TM	0.81362	0.29566	0.82971

Table 9.10 Energy fuzzy partition quality (2 labels)

Partition	Partition coefficient(max)	Partition entropy(min)	Chen index(max)
HFP	0.78374	0.31852	0.66240
Regular	0.52352	0.66896	0.17696
K-means	0.77454	0.33891	0.67045

The most important part of the knowledge-based systems is the reasoning mechanism that induces decision rules. Since the fuzzy partition is at the core of induction methods in fuzzy logic, we proposed the initialisation of the fuzzy partitions by two approaches: first, a purely automatic method of induction (*k-means, HFP* and *regular*) and, second, the approach resulting from information extraction from textual sources, as discussed in Sect. 3, to compare between the different methods. Subsequently, we have established linguistic terms to build the rules and modal point. Note that the shape of the partition function is to be discussed with experts.

We analysed the results using the different indices partition, and we identified the features with QRS and RRs; partition coefficient (*max*), partition entropy (*min*) and Chen index are better compared to other algorithms (k-means, HFP and regular). Since the available data only covers part of the real interval, the two approaches have been integrated, to provide a partition which is both accurate and interpretable.

9.6 Rule Base Generation

Induction is the process of generating rules from data. Its aim is to produce general statements, expressed as fuzzy rules in our case, valid for the whole set, from partial observations. The observed output, for each sample item, is part of the training set allowing supervised training procedures. Many methods are available within the fuzzy logic literature [31], but only those which generate rules sharing the same fuzzy sets can be applied in this context [33]. Thus, we have chosen the methods, which are implemented in FisPro [31], and they are used by KBCT [2], as they can be run with previously defined partitioning.

9.6.1 Knowledge Base Accuracy

In order to obtain an accuracy measure, we compared the inferred output with the observed one in a real system. In classification systems, the most common index is defined as the number of misclassified cases. We only considered the following indices:

- Unclassified cases (*UC*): Number of cases from data set that do not fire at least one rule with a certain degree.
- Ambiguity cases (*AC*): Number of remaining cases for which the difference between the two highest-output confidence levels is smaller than an established threshold (*AmbThres*). More specifically, we also have:

  - *AC* (*total*): All detected ambiguity cases
  - *AC* (*error*): Only those ambiguity cases related to error cases (observed and inferred outputs are different)

– *EC*: Error cases. Number of remaining cases for which the observed and inferred output classes are different

• *Data* (*total*): The total number of instances in the data set.
• Error cases (*EC*): Number of remaining cases for which observed and inferred values are different.

An efficient KB should minimise all of them by offering an accurate (reducing EC), consistent (reducing AC) and complete (reducing UC) set of rules. They can be combined to define the next accuracy index (Fig. 9.10, Table 9.11):

$$Accuracy = 1 - \frac{EC + AC(Error) + UC}{DATA(TOTAL)}$$

$$Accuracy(CONS) = 1 - \frac{EC + AC(TOTAL) + UC}{DATA(TOTAL)}$$

$$Accuracy\,(BT = 0) = 1 - \frac{EC + AC(Error) + UC(Error)}{DATA(TOTAL)}$$

9.7 Evaluation

The method discussed in this chapter was evaluated by considering a variety of experimentations. We first carried out the fuzzy partition, and subsequently we induced the corresponding decision rules and calculated and assessed the quality criteria to measure the accuracy of each approach. We noted that the best results are FDT1, FDT4, FDT7, FDT14 and FDT15, as shown in Table 9.10 with the following parameters:

• Coverage (cv%): percentage of data samples from the selected data set that fire at least one rule in the rule base with an activation degree higher than the predefined blank threshold (BT)
• Accuracy (ac): percentage of data samples properly classified
• Accuracy (*acons*): percentage of data samples properly classified
• Average confidence firing degree (*acfd*): mean value of the firing degree related to the winner rule for the whole data set
• Minimum confidence firing degree (*micfd*): minimum value of the firing degree related to the winner rule for the whole data set
• Maximum confidence firing degree (*macfd*): maximum value of the firing degree related to the winner rule for the whole data set

Rules

Rule	Type	Active	IF RR0	AND RRs	AND COMP	AND QRS	AND PP	AND ENERGIE	THEN Class
1	1	yes			Late L				1.0
2	1	yes		NOT(Irregular R)	NOT(Late L)	Small			1.0
3	1	yes	Irregular L		NOT(Late L)	Average			1.0
4	1	yes	NOT(Irregular L)			Average			1.0
5	1	yes			Late R	Large	Small		1.0
6	1	yes		Irregular R	NOT(Late L)	Small			3.0
7	1	yes			Regular	Large	Small		2.0
8	1	yes			NOT(Late L)	Large	Average		2.0
9	1	yes			NOT(Late L)	Large	Tall		2.0
10	E	yes	Irregular L	Irregular R	Late R	Large	Tall	high	2.0
11	E	yes	Irregular R	Irregular L	Late L	Average	Tall	high	2.0

Fig. 9.9 Rules: FDT14 (expert and text mining) and fuzzy decision tree algorithm

Table 9.11 Quality measurements

KB	cv %	ac	acons	abt = 0	acfd	micfd	macfd	me	msce
FDT1	100	0.914	0.908	0.914	0.865	0	1	3	0.007
FDT4	100	0.424	0.303	0.424	0.516	0	1	3	0.125
FDT7	100	0.873	0.87	0.873	0.448	0	0.842	3	0.054
FDT15	100	0.717	0.704	0.717	0.704	0.013	1	3	0.013
FDT14	100	0.926	0.921	0.926	0.903	0	1	3	0.006

- Max error (me): maximum difference between the observed class and the inferred one
- Mean square classification error ($msce$)

Furthermore, FDT14, which was generated by attributes resulting from the text mining extraction (RRs and QRS) integrated with k-means and HFP algorithms, produced the best result with a classification rate of 92.6 % and a 0.16 interpretability value. On the other hand, FDT14, created by expert fuzzy partition with fuzzy decision tree induction algorithm, gave a lower classification rate and interpretability value of 71.7 % and 0.025, respectively.

While considering FDT1, FDT4 and FDT7 [2], we noticed that the interpretability value was zero, which is clearly explained by the very large number of rules.

9.8 Conclusion

In this chapter a case study focusing on information extraction within the medical domain is discussed. The use of fuzzy logic as a platform for communication between the different sources of knowledge is a successful solution to manage the fusion of knowledge in a database of common rules. The contribution of text mining in the extraction of knowledge from a large unstructured data set provided excellent results, with an accuracy of 92.6 %, compared to the other algorithm which may present high accuracy but lacking in interpretability. Furthermore, this approach offers more flexibility and transparency, with a contribution of the expert who is involved in the decision-making process, clearly showing the applicability and potential of this method.

References

1. World Health Report 2013. Retrieved Oct 06 2014, from http://www.who.int/whr/en/
2. Behadada O, Trovati M, Chikh MA, Bessis N (2015) Big data-based extraction of fuzzy partition rules for heart arrhythmia detection: a semi-automated approach. Concurr Comput Pract Exp. doi:10.1002/cpe.3428

3. Alonso JM, Luis Magdalena L (2009) An experimental study on the interpretability of fuzzy systems. In: Proceedings of the joint 2009 international fuzzy systems association world congress and 2009 European society of fuzzy logic and technology conference, Lisbon, 20–24 Jul 2009
4. Gacto MJ, Alcala R, Herrera F (2011) Interpretability of linguistic fuzzy rule-based systems: an overview of interpretability measures. Inform Sci 181(20):4340–4360. doi:10.1016/j.ins.2011.02.021
5. Quinlan JR (1990) Learning logical definition from relation. Mach Learn 5:239–266, http://dx.doi.org/10.1007/BF00117105
6. Alpaydin E (2010) Introduction to machine learning. MIT Press, Cambridge, MA
7. Pazzani M, Kibler D (1992) The utility of knowledge in inductive learning. Mach Learn 9(1):57–94, http://dx.doi.org/10.1007/BF00993254
8. Rumelhart D, Hinton G, Willams R (1986) Learning internal representations by error propagation. In D. Rumelhart, J. McClemlland (eds) Parallel distribution proceeding: exploration in the microstructure of cognition, Foundations, vol 1. MIT Press, Cambridge, MA, pp 318–362
9. Meau YP, Ibrahim F, Naroinasamy SAL, Omar R (2006) Intelligent classification of electrocardiogram (ECG) signal using extended Kalman filter 441 (EKF) based neuro fuzzy system. Comput Methods Progr Biomed 82(2):157–168. doi:10.1016/j.cmpb.2006.03.003
10. Piotrkiewicz M, Kudina L, Mierzejewska J, Jakubiec M, Hausmanowa-Petrusewicz I (2007) Age-related change in duration of after hyperpolarization of human motoneurones. J Physiol 585:483–490
11. Soheilykhah S, Sheikhani A, Sharif AG, Daevaeiha MM (2013) Localization of premature ventricular contraction foci in normal individuals based on multichannel electrocardiogram signals processing. SpringerPlus 2:486
12. Quinlan JR (1993) C4.5: programs for machine learning. Morgan Kaufmann, Los Altos
13. Hosseini HG, Luo D, Reynolds KJ (2006) The comparison of different feed forward neural network architectures for ECG signal diagnosis. Med Eng Phys 28(4):372–378
14. Chiang DA, Chow LR, Wang YF (2000) Mining time series data by a fuzzy linguistic summary system. Fuzzy Sets Syst 112:419–432
15. Pedrycz W (1998) Fuzzy set technology in knowledge discovery. Fuzzy Sets Syst 98:279–290
16. Zadeh LA (1965) Fuzzy sets. Inf Control 8:338–353
17. Yu S, Chou T (2008) Integration of independent component analysis and neural networks for ECG beat classification. Expert Syst Appl 34(4):2841–2846
18. Krishnapuram R, Joshi A, Nasraoui O, Yi L (2001) Low complexity fuzzy relational clustering algorithms for web mining. IEEE Trans Fuzzy Syst 9:595–607
19. Frigui H, Krishnapuram R (1999) A robust competitive clustering algorithm with application in computer vision. IEEE Trans Pattern Anal Mach Intell 21(1):450–465
20. Baldwin JF (1996) Knowledge from data using fuzzy methods. Pattern Recognit Lett 17:593–600
21. Manning CD (1999) Foundations of statistical natural language processing. MIT Press, Cambridge, MA
22. Liu B (2012) Sentiment analysis and opinion mining. Morgan & Claypool Publishers, San Rafael
23. de Marneffe MF, MacCartney B, Manning CD (2006) Generating typed dependency parses from phrase structure parses. In: Proceedings of LREC-06, Genoa
24. PubMed. Retrieved 06 Oct 2014, from http://www.ncbi.nlm.nih.gov/pubmed/
25. Duda R, Peter Hart E (1973) Pattern classification and science analysis. John Wiley and Sons, New York
26. Moody GB, Mark RG (2001) The impact of the MIT-BIH arrhythmia database. IEEE Eng Med Biol 20(3):45–50
27. Pan J, Tompkins WJ (1985) A real-time QRS detection algorithm. IEEE Trans Bio-Med Eng 32(3):230–236

28. Schönbauer R, Sommers P, Misfeld M, Dinov B, Fiedler F, Huo Y, Arya A (2013) Relevant ventricular septal defect caused by steam pop during ablation of premature ventricular contraction. Circulation 127(24):843–844
29. Yager RR (1996) Database discovery using fuzzy sets. Int J Intell Syst 11:691–712
30. Casillas J (2003) Accuracy improvements in linguistic fuzzy modelling, vol 129, Studies in fuzziness and soft computing. Springer Science & Business Media, Heidelberg, pp 3–24
31. Guillaume S, Charnomordic B (2012) Fuzzy inference systems: an integrated modelling environment for collaboration between expert knowledge and data using FisPro. Expert Syst Appl 39(10):8744–8755
32. Chen T (2013) An effective fuzzy collaborative forecasting approach for predicting the job cycle time in wafer fabrication. Comput Ind Eng 66(4):834–848
33. Wei Q, Chen G (1999) Mining generalized association rules with fuzzy taxonomic structures. Proceedings of NAFIPS 99, IEEE Press, New York, pp 477–481

Chapter 10
A Platform for Analytics on Social Networks Derived from Organisational Calendar Data

Dominic Davies-Tagg, Ashiq Anjum, and Richard Hill

Abstract In this paper, we present a social network analytics platform with a NoSQL Graph datastore. The platform was developed for facilitating communication, management of interactions and the boosting of social capital in large organisations. As with the majority of social software, our platform requires a large scale of data to be consumed, processed and exploited for the generation of its automated social networks. The platforms purpose is to reduce the cost and effort attributed to managing and maintaining communication strategies within an organisation through the analytics performed upon social networks generated from existing data. The following paper focuses on the process of acquiring and processing redundant calendar data available to all organisations and processing it into a social network that can be analysed.

10.1 Introduction

The success and performance of large organisations is highly dependent on communication at every level of an organisation. The ability to react to change proves much more efficient when the appropriate social structures and communication strategies promote the dissemination of knowledge and collaboration. With the coming of large social networking sites such as Facebook or Twitter, many organisations have taken an interest in such services' ability to share and pass along knowledge, thus resulting in various internal social network implementations or just the addition of some social functionality (blogs, forums, wikis) being added to existing services.

The first problem the proposed platform will address is that of forcing social network functionality onto employees; these internal implementations are rarely adopted and commonly forgotten soon after implementation [1]. Instead this platform will repurpose redundant data widely available to all organisations (and

D. Davies-Tagg (✉) • A. Anjum • R. Hill
Department of Computing and Mathematics, University of Derby, Derby, UK
e-mail: dominicdaviestagg@gmail.com; A.Anjum@derby.ac.uk; r.hill@derby.ac.uk

© Springer International Publishing Switzerland 2015
M. Trovati et al. (eds.), *Big-Data Analytics and Cloud Computing*,
DOI 10.1007/978-3-319-25313-8_10

most individuals) and use it to create a complex social network structure with minimal user input, the data in question being calendar data.

The second item that the platform aims to deliver is a tool that facilitates the analysis of calendar-generated social networks to aid in decision-making about the existing communication structure within the organisation, so individuals and groups that are poorly connected to the network can be identified and have communication strategies put in place to facilitate better communication or identification of highly connected individuals who the communication structure is too dependent on [2].

The rest of this paper is organised as follows. In Sect. 10.2, we discuss the principles and research relevant content that the project is built from. Section 10.3 outlines the proposed platform, providing an overview of how the data is collected and processed by the platform. A detailed breakdown of the specific technologies that the platform will use is discussed in depth in Sects. 10.4 and 10.5 which focuses on the analytical processes and the knowledge that is gained from such a platform. Finally the paper concludes with noting the current merits of the platform and the future direction for it.

10.2 Literature Review/Related Work

In this section, an overview of existing approaches and tool is discussed.

10.2.1 Social Capital and the Exchange of Knowledge/Resources

Social capital can be defined as a resource that is produced by relationships between people [3]. The definition extends beyond individual people and can apply to groups and even organisations. Simply put by seeking out relationships with different people, we get access to them and their own social structures as a resource [4], a resource being knowledge, skills, position or even connection to others. The advent of social networking sites has facilitated further adoption of social capital as they provide the ability to easily manage connections while also providing a much larger network to connect with and use as a resource [5]. There are various complexities and definitions if we looked deeper into social capital, especially where organisational social capital is concerned, but for the context of this platform, the simple definition above will suffice.

We can think of social capital as a point system with those with a high number of connections to others as having a higher social capital, the types of connections themselves also matter; being connected to others with a high social capital will be of greater benefit than a connection with an unconnected individual. The individual could prove a useful resource themselves, but the resource pool they have access to will be much more limited than someone who is highly connected.

Dissemination of knowledge within an organisation is largely dictated by organisational structure, but the social network structure and individual relationships that people have are instead based on trust and reciprocity; just communicating within the confines of dictated organisational structure results in only knowing what everyone else in your defined workgroup knows [6]. Making connections and nourishing external relationships provide access to a much greater wealth of knowledge, and these types of connection are a great source for innovation and collaboration with others, which is essential for any organisation that wants to remain agile and competitive [7].

10.2.2 Social Capital and the Exchange of Knowledge/Resources

The study of organisational social networks is not a new concept, with Monge and Contractor taking social network analysis concepts and applying them to study patterns of communication to discern how information flows across and between organisations [8]. Within organisations, there was also a shift in the past decade going from "what you know?" being most important to "Who you know?" holding just as much weight, with activities like networking (creating connections, maintaining connections and expanding one's own network) becoming common requirements of many people's jobs [6].

The majority of research on social networks at least within the context of organisations is that of identifying how people, groups or whole organisations communicate; from this, potential problems and gaps in communication can be identified and resolved [9].

When analysing organisational social networks, there are certain patterns we can identify that tell us a lot about an organisation's network strengths and weaknesses; these are centrality, outliers, subgroups and even the direction that communication flows [10]. Centrality is a focus on the most prominent people in a network; this is quantified by the amount of connections a person has and the strength of these connections. Someone with high centrality will most likely be a key communicator and influence a lot of others, but if this person is removed, the organisation can potentially be adversely affected. Peripheral people or outliers are the people on the edges of the network that are very loosely connected; there could be organisational reasons as to why these individuals are not so highly connected, but isolation and non-inclusion in the network are bad for communication but also result in under usage of that person as a resource. Subgroups will occur naturally, be it the subgroup denoted by organisational grouping or a specific project or even based on location; by breaking networks into subgroupings, we can then understand how different groups interact as a whole instead of the individuals that comprise the group. Finally within network structure, it is important to understand the direction that connections flow; this can tell us who is doing the communicating and who is being communicated with.

10.2.3 Repurposing Redundant Organisational Data

There is always going to be more old data than new data, but old data can still be useful; it may have served its initial purpose, but in another's hands, additional new value can be created [11]. The reuse of data can come in different forms such as taking the exact data set, but applying new knowledge to the data to derive different information, it could also be the collation of the data with other data to provide a much larger resource to analyse and even taking the data and applying a totally different purpose to it altogether. In data science, the whole point of repurposing data is with the goal of generating value where none is expected [12].

Exposing social networks in existing data isn't a new concept, and arguably social networks can be derived from anytime two people have some form of interaction, but some data might be more relevant than others similar to this project's intended usage of calendar data; in a similar vein, prior research has been done regarding the extraction of social networks from organisational email [13], [14].

10.2.4 Graph Databases

NoSQL or better known now as "Not only SQL" stems from the ideas that RDBMS (relational database management systems) are not the end-all solution to our data storage requirements and that other types of database could perform better depending on the situation. NoSQL's rise can be attributed to large Web 2.0 (Amazon, Google, Facebook, etc.) companies reaching the limits of RDBMS and creating their own databases that would scale horizontally and provide necessary levels of availability by ignoring common relational database practices such as consistency and strict data schemas. NoSQL is not just about the hugely scalable, available implementations; it's an umbrella term used to describe any non-relational database, and so NoSQL can be split up into a variety of different paradigms such as column, document, key/value store, graph, object and grid each with a variety of different characteristics.

Graph databases are comprised of nodes, edges and properties. Nodes essentially being objects and edges being the connections and structural relationships between these objects, and properties are the fields and values associated with both the nodes and edges. Graph databases are based upon graph theory that dates back to a paper in 1935 by a mathematician/physicist known as Leonhard Euler; graph theory is basically the study of graphs [15].

When researching graph databases, the most prominent example of usage is with social networking data [16, 17]. This is because the storing of the connections between entities using graph theory accurately maps to the storage of stakeholders and relationships in social networking data. You could argue that

RDBMS also store the relationships between entities, but the tabular data structure of a relational database is not suited to the traversal and querying often required of highly related data, unlike a graph database which has the data structure and commonly a selection of pre-existing tools to facilitate graph traversals and queries [18].

10.3   Proposed Platform

We propose a web application that sources calendar data to create complex social network visualisations that can be analysed for the identification of failings in communication and also immediate problems with how specific individuals or groups interact with each other.

The solution can be broken down into a five-step process that will be expanded upon within the following section. Each step will be explored in depth providing evidence and reasoning behind its individual importance and as part of the process as a whole. Figure 10.1 provides a visual representation of the steps and the order in which they occur. Arguments could be made for the combination of certain steps, but for this discussion, the individual steps merit enough discussion on their own.

10.3.1   Capture: Capturing the Calendar Data

The base data for the platform will be calendar data, nothing specialised just standard calendared meetings extracted from people's calendars. Calendar data provides a variety of information, a time and date for the meeting, log of all attendees, context for the meeting and potentially even a location. The person

Fig. 10.1 Simplified platform process diagram

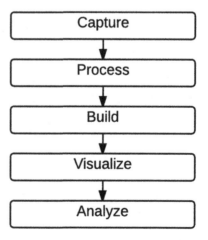

driving these meetings can also be inferred by who initiated the meeting invite. All of these elements are important data items that add further perspective to each interaction and so require capturing for each individual meeting in a person's calendar.

Extracting data from calendars comes with the advantages of the data being strictly enforced by time and date allowing for targeted ranges of data to be easily captured.

10.3.2 Process: Processing the Captured Data into Social Data

Calendar data isn't similar to social data, so it requires transforming into a format that resembles an interaction between two people. Instead of an attached listing of attendees of a meeting, it has to be modelled as an attendee connecting with another attendee, so every single attendee (including the organiser) will have a connection with every other individual person. Figure 10.2 is a simplistic representation of the transformation that occurs on the collected meeting data in the process step; this is explained below with the simple visuals found in Figs. 10.3 and 10.4.

The processing algorithm starts by extracting all the data it can from the input meeting data, data such as meeting names, dates and times that are stored within a dedicated meeting node, and then all the meeting attendee data is captured but also the information about this specific attendee attending this meeting is also captured. Figure 10.3 represents a single meeting with four attendees, in which each attendee

```
Input: Meeting Data
    BEGIN:
        CREATE Meeting Object
        FOREACH Meeting Attendee
            CREATE Attendee Object
            CREATE Meeting --> Attendee Connection Object
        UNTIL no more meeting attendees remain

        FOREACH Attendee Object (Outer)
            FOREACH Attendee Object (Inner)
                IF Attendee Object (Outer) EQUAL TO Attendee Object(Inner)
                    THEN
                        // Do Nothing
                    ELSE
                        CREATE Attendee(Outer) --> Attendee(Inner) Connection Object
                UNTIL no more attendee objects remain
            UNTIL no more attendee objects remain
    END:
Output: Meeting Object,
        Attendee Object(s),
        Meeting --> Attendee Connection Object(s)
        Attendee_Connection Object(s)
```

Fig. 10.2 Process pseudocode algorithm

Fig. 10.3 Simple meeting and attendees visual

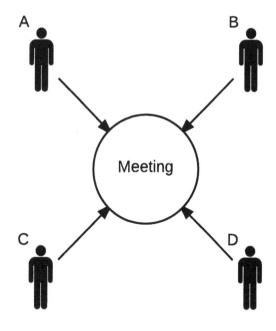

Fig. 10.4 Breakdown of meeting interactions

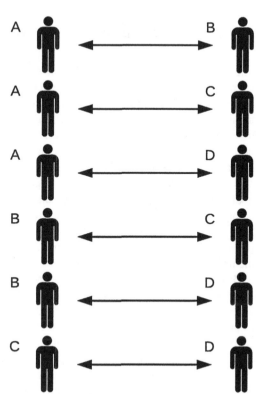

is going to the meeting, and this is pictured as an arrow from the attendee towards the meeting.

Capturing the data as is could work very similarly to our planned implementation except that instead of taking a simplistic path of a person node connected to another person node via a single edge, an additional node and edge would be required to include the meeting and both person nodes interacting with the meeting (node, edge, node, edge and node).

Instead, we ignore the meeting itself and focus on the direct interactions between each individual, resulting in data formatted simply as node, edge and node. In Fig. 10.2, this is represented as the nested for each loop that iterates over the collection of attendee objects in both the inner and outer loops and captures that they are all connected by this meeting interaction. Figure 10.4 depicts the simple meeting in Fig. 10.3 but broken down into a simplistic representation of a bidirectional interaction between every attendee. Arguably this produces more edges that need creating but drastically simplifies the data structure and reduces the number of hops required to extract or query the exact same interaction data we are interested in.

The meeting data structure in Fig. 10.3 is still captured within the application (and is still present in Fig. 10.2) as it can be used for other knowledge extraction/analysis outside the scope of this paper.

10.3.3   Build: Building the Social Network

The structure of a network is that of nodes and edges; in a social network, these are placeholders for people and the interactions between them. The previous step of processing deconstructed the calendar data into a simplistic format that allows for thousands of these connections to be brought together to form a highly connected network of interpersonal interactions. The pseudocode presented in Fig. 10.5 is how the data that was harvested in the previous step from simple meeting data is constructed into a complex network graph structure.

Taking the data from the process stage (Fig. 10.2), converting it into simple nodes and edges and combining them all create a simple graph of interactions of a single meeting as seen in Fig. 10.6. From this we can easily discern that node A is connected to B, C and D. However, just knowing that an interaction occurs provides little information about the strength/frequency of the relationship between the nodes.

The secondary function of the build step is to calculate the strength of connections between two nodes, and this occurs in Fig. 10.5 towards the end of the loop when we are sure our attendee nodes have been created. Figure 10.4 depicted a list of interactions for a single meeting, but from a calendar thousands of such data sets will be produced. Some of this data will be interacting with existing connections over and over again.

Instead of discarding repeated interactions, a value is instead added to the edge connecting the two nodes, and this numerical value is incremented for each

```
Input:   Attendees Object(s),
         Attendee_Connection Object(s)
     BEGIN:
         FOREACH Attendee_Connection Object
             IF Attendee_Connection.SourceAttendee NODE EXISTS in DB
                 then
                     // do nothing
                 else
                     CREATE/STORE Attendee NODE (Attendee_Connection.SourceAttendee)

             IF Attendee Connection.TargetAttendee NODE EXISTS in DB
                 then
                     // do nothing
                 else
                     CREATE/STORE Attendee NODE (Attendee_Connection.TargetAttendee)

             IF Attendee_Connection EDGE EXISTS in DB
                 then
                     INCREMENT EDGE value Attendee_Connection.edgeWeight by 1
                 else
                     CREATE/STORE Attendee_Connection EDGE
         UNTIL no more meeting attendee_Connection objects remain
     END:
Output: Attendee Node(s),
         Attendee Connection Edge(s)
```

Fig. 10.5 Build pseudocode algorithm

Fig. 10.6 Simple graph of
nodes and edges

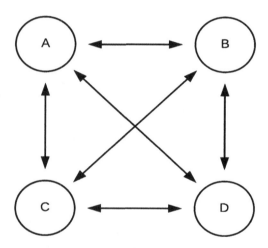

interaction between nodes that is encountered. In graph theory, this is known as a weighted graph [19]; the weighting itself can represent anything, but in our instance, it represents the total interactions between two nodes which we then interpret as the level of connectivity between two people as is demonstrated in Fig. 10.7.

Figure 10.8 is a perfectly reasonable graph structure combining and representing what occurs if all the simple data sets that are created from the process step are combined, and without question it looks messy and convoluted. Opting instead for a weighted graph (Fig. 10.7) provides a much simpler and visually appealing data structure that will prove easier to query and analyse.

Fig. 10.7 Weighted graph

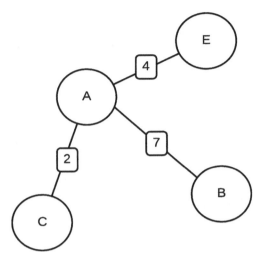

Fig. 10.8 Edge-dense graph
visual

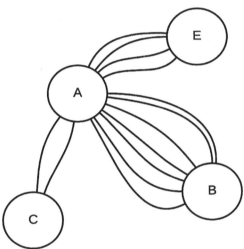

10.3.4   *Visualise: Visualising the Social Network Structure*

The visualisation step is a necessary requirement because of the large amount of
data that will be collected and the resulting complexity of the highly correlated data.
Small graphs such as in Fig. 10.8 can be easily represented in text and evaluated,
but when the amount of data scales to thousands of edges and nodes, it is no longer
a viable option such as the huge social graphs that can be created from Facebook
or Twitter data (Fig. 10.9). Instead, we visualise the data as a graph of our person
nodes and interaction edges, and such visualisation allows for the identification of
patterns in data and a perspective that allows humans to analyse huge amounts of
data.

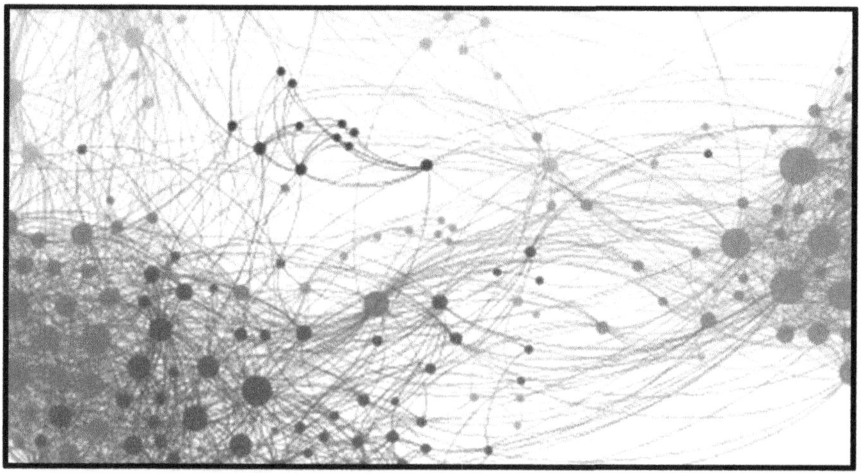

Fig. 10.9 Dense Facebook social graph

Visualisations also act as a great platform for facilitating interactions with data by providing functions to manipulate, zoom, pan and inspect details of a graph you provide a human-centric interface that promotes exploration of the structure and patterns of the data.

10.3.5 Analyse: Performing Analysis Against the Social Network

The four prior steps discussed above captured and transformed calendar data into a highly complex weighted graph that is delivered in a human-centric format, ready for analysis to be conducted and knowledge produced.

This step of the process is more reliant on the user's interaction; the platform provides the data and visualises it alongside augmenting it with other details and interactive functions, but it is the user who actively explores the data for specific knowledge making use of the platform and its tools.

Specific examples and an in-depth breakdown of the analysis that the platform facilitates are found in the results section.

10.3.6 Experimental Setup

The following section focuses on the individual components that comprise our proposed solution (Fig. 10.10) that can simply be broken down into three separate components.

Fig. 10.10 Solution
overview diagram

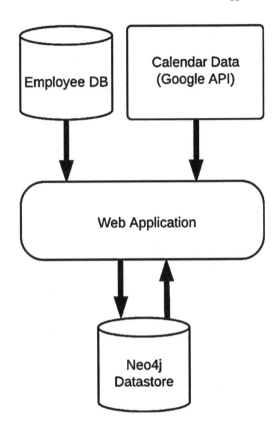

- External data sources harvested for data (employee DB, Google Calendar API).
- Web application that processes data into information and provides an interface to interact with this data.
- Graph datastore that allows for simple and efficient storage/retrieval of data alongside complex graph-based queries against our data.

10.3.7 Solution Setup

Google APIs and OAUTH2 Authentication

Calendar data provides the primary interactions that occur within our system, giving information such as who a person interacted with at a specific time and place and even the meetings' topic. Calendar data provides a great number of interactions that will populate a network of people but only with a limited information and often just an email to identify who attended a meeting, so other data sources are needed to complete a useable profile. Calendar data is sourced from Google Calendar, and so the Google Calendar API was used to access the data, utilising OAUTH2 authentication.

Employee Database

HR data is essentially a collection of profiles that exist for each individual belonging to an organisation, providing a wide array of data about a person that is just as relevant for a profile on a social network.

JDBC is used to request the HR data that is stored within a MS SQL database, alongside employee data; other data such as organisational structure is available telling us who is in charge of what departments and organisational groups providing network context about how people are grouped. HR data only provides profile data for employees; this will mean that profiles for external stakeholders will remain largely incomplete unless someone created a profile information for them manually.

Neo4j Graph Database

The graph database of choice is Neo4j, a java-based solution produced by Neo Technologies. It is presently the most popular and widely used graph database [20] touting customers such as HP, Adobe and Cisco [21]. Neo4j is open source (GPL), but unlike alternative graph database solutions, it also offers commercial licensing and support. Features include ACID compliance, C++ Core and a wide support for a variety of APIs (Cypher, Blueprints and Native Java APIs) making it versatile to develop with.

Neo4j provides a simple almost SQL-like query language known as Cypher that allows for effective preparation of queries against the stored data. This offers unique set of keywords and commands that facilitate graph-based operations such as traversal and path finding techniques. Alternative graph databases often require much more complex, hard-coded queries that can prove a challenge to implement and maintain. Like a lot of other NoSQL implementations, Neo4j is schemaless and is ideal for the incomplete and partial records that the proposed platform will accumulate as it builds profiles and knowledge to build the social network.

Spring MVC Web Application

The web application was built in Java and utilises the Spring Framework, chosen due to its compatibility with Neo4j and the variety of ease of use functionalities provided within Spring and its associated modules. Once completed, the application was deployed to and ran from an Apache Tomcat web server, running from the same machine that hosted the Neo4j data service to provide optimal speed and performance in communication between the two.

The browser facing front end is provided by Twitter Bootstrap, with complex data visualisations provided by the D3.js JavaScript library. D3.js is a powerful library that allows for various types of data visualisations (bar charts, pie, trees, choropleth maps, etc.), our application makes heavy usage of visualisations such as the force-directed graphs that are appropriate for social network visualisations.

10.3.8 Hardware Setup

The project was run from a machine running Windows 8 with an i5-3570 K CPU @ 3.40 Ghz, 16 GB of RAM and utilising a 256 GB SSD for the storage of the entire solution. Depending on the amount of data handled and performance expectations, a more or less powerful machine could easily be used. The main performance hurdle will be preparation and ingestion of the data, so large data sets would benefit from a higher performance system.

10.4 Results

The tool is still in development, so the following section provides preliminary statistics captured and an in-depth analysis of the achieved functionalities and benefits of the current implementation, and traditional quantitative/qualitative means of evaluation will follow if further research and development is appropriate.

10.4.1 Outlier Detection

Outliers can be defined as data that does not fit the structural norm and instead sticks out as being different; for our social network-based definition, we will consider those loosely connected nodes on the edge of the network as outliers. Outliers can be detected either via sight when someone inspected the network visually and notices the deviance within the network structure but also can be identified further by queries that pick out the nodes that fell below a certain threshold of connections as demonstrated in Fig. 10.11. Finding these outliers and bringing them into the network will result in these members of the network being utilised as a much more effective resource; it will also improve communication flow by providing a greater chance of communicating with others.

10.4.2 Detection of Outlying Groups

Similar to the isolated individuals, Fig. 10.12 depicts a group that is isolated from the network as a whole and exists on the periphery of the network, and the same problems in communication can occur and be more prevalent if a group is only interacting with itself. Identification of this is most easily done visually, but queries similar to the first can be tailored so it aggregates the individuals that make up the group.

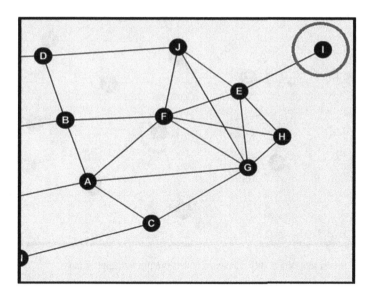

Fig. 10.11 Screenshot of outlier detection from the web application

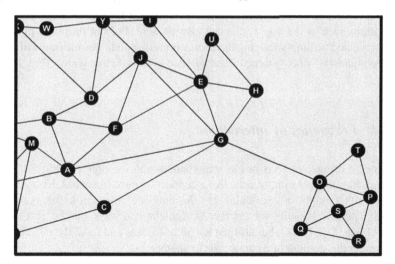

Fig. 10.12 Screenshot of isolated group detection from the web application

10.4.3   Identification of Key Communicators for Specific Groups and Highly Connected Individuals

Those densely connected and central to a network have high importance, and by utilising their connections, they can effectively drive change and innovation throughout a network [22]. Identification of such individuals can be done in a

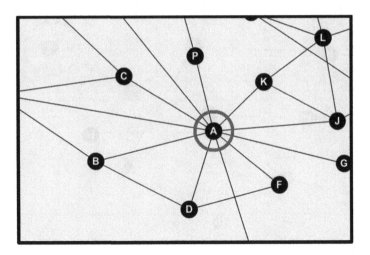

Fig. 10.13 Screenshot of a highly connected node from the web application

variety of ways such as a simple count of direct connections and an evaluation of the connections that the other nodes are connected to; various algorithms and calculations such as the eigenvalue can also be used. For our proposed platform, the first method of simply tallying the connections of a node was utilised with those with the highest tally being targeted and labelled as key communicators (Fig. 10.13).

10.4.4 Frequency of Interaction

The amount of interaction is just as important as who we interact with; the higher the frequency, the more importance that a connection most likely has. Every edge is denoted with a weight value calculated by the number of interactions that occur with a specific person. Visually we can represent this in two ways, one being a thicker width of line (Fig. 10.14), but also the length of the line and force the edge has can also present the strength of a connection to another.

10.4.5 Experiment Data Statistics

The following section provides a selection of statistics (Tables 10.1 and 10.2) that are not part of the platform but were instead queried from the database the platform uses; the data used was captured from 10 people using the platform within a range of 4 months.

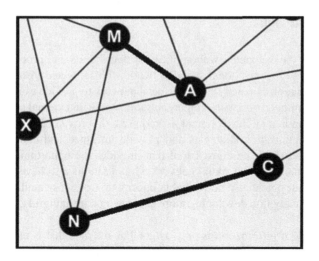

Fig. 10.14 Screenshot of edge weight being visualised

Table 10.1 Platform collected data overview

Unique user nodes	164
Unique interaction edges	1109
Meetings captured	426

Table 10.2 Meeting analysis data

Average meeting size (mean)	5.2
Average edge weight (mean)	2.7
Average meeting length (minutes)	47

Table 10.1's Unique User Nodes tells us that from the 10 people's calendars used over the 4-month period, 164 unique meeting attendees were encountered, 174 if counting the existing 10 platform users. Throughout these 174 meeting attendees and 426 meetings, 1109 interaction edges were recorded, and these are unique and only count the initial meeting; subsequent encounters after the first are excluded from this count. This data provides an insight into the amount of data collected by the platform but adds little knowledge that would be of benefit to an organisation.

Table 10.2 depicts statistical values that would be of interest when performing meeting analysis, and no additional work or effort was required to produce these results, but just simple queries performed against the graph database. This is a simple repurposing of data we are collecting anyway to extract meaningful knowledge about meetings that occur. The meeting data can also contribute towards improving communication within an organisation, and details about meetings being oversized or running too long can be taken into consideration, and changes can be implemented to better serve an organisation need [23].

10.5 Conclusions

We successfully developed a web application that ingests and processes calendar data into a large organisational social network with tools and visualisations that allow for interaction and analysis of the social network to provide valuable information about communication within an organisation and with external stakeholders.

More research and development is required to provide functionalities that facilitate communication strategies and provide informed and computed recommendations instead of the current tool that provides the opportunity for humans to identify elements of the social network. Organisations also have other sources of redundant but potentially networkable information such as emails or blogs that could factor greatly into developing more accurate and meaningful communication strategies.

The validity of information is only as good as the data that is used to produce that information; this also proves true of the calendar data we utilised, so a deeper investigation needs to be taken into how accurately a network generated from calendar information mirrors the reality of an organisation's social network. Further research would take a qualitative approach with questionnaires taken to corroborate the network structures produced.

References

1. Mergel I, Muger G, Jarrahi MH (2012) Forming and norming social media adoption in the corporate sector. 2012 iConference. ACM, Toronto/New York, ACM, pp 152–159, 7 Sept 2012
2. de Paula, RA, Appel A, Pinhanez CS, Cavalcante VF, Andrade CS (2012) Using social analytics for studying work-networks: a novel, initial approach. In: Proceedings of the 2012 Brazilian symposium on collaborative systems. IEEE Computer Society, Sao Paulo/Washington, DC, pp 146–153, 15 Oct 2012
3. Coleman JS (1988) Social capital in the creation of human capital. Am J Sociol 94:95–120
4. Sander T, Phoey-Lee T (2014) Determining the indicators of social capital theory to social network sites. In: User science and engineering. IEEE, Shah Alam/Washington, DC, pp 264–268, 2–5 Sept 2014
5. Brandtzæg PB (2012) Social networking sites: their users and social implications – a longitudinal study. J Comput Mediat Commun 17(7):467–488
6. Crona B (2006) What you know is who you know? Communication patterns among resource users as a prerequisite for co-management. Ecol Soc 11(7). http://www.ecologyandsociety.org/vol11/iss2/art7/. Accessed 20 Jan 2015
7. Fanfan Y (2011) Insights into social network theory and entrepreneurship. In: 2011 international conference on management and service science (MASS). IEEE, Wuhan/Washington, DC, pp 1–4
8. Inkpen AC, Tsang EWK (2005) Social capital, networks, and knowledge transfer. Acad Manag Rev 30(1):146–165
9. Cornelissen J (2011) Corporate communication: a guide to theory and practice, 3rd edn. Sage, London
10. Hanneman RA, Riddle M (2005) Introduction to social network methods. University of California, Riverside. http://www.faculty.ucr.edu/~hanneman/nettext/. Accessed 20 Jan 2015

11. McKay D (2014) Bend me, shape me: a practical experience of repurposing research data. In: 2014 IEEE/ACM joint conference on digital libraries (JCDL). IEEE, London/Washington, DC, pp 399–402, 8–12 Sept 2014
12. Wallis JC (2013) If we share data, will anyone use them? Data sharing and reuse in the long tail of science and technology. PLoS One 8(7):1
13. Bird C, Gourley A, Devanbu P, Gertz M, Swaminathan A (2006) Mining email social networks. In: Proceedings of the 2006 international workshop on mining software repositories (MSR '06). ACM, New York, pp 137–143
14. Shetty J, Adibi J (2005) Discovering important nodes through graph entropy the case of Enron email database. In: Proceedings of the 3rd international workshop on link discovery, Chicago, pp 74–81, 21–25 Aug 2005
15. Wilson RJ (2012) Introduction to graph theory, 5th edn. Pearson Education Limited, Southport
16. Li-Yung Ho, Jan-Jan Wu, Pangfeng Liu (2012) Distributed graph database for large-scale social computing. In: 2012 IEEE 5th international conference on cloud computing (CLOUD), pp 455–462, 24–29 June 2012. Honolulu, HI, USA
17. Fan W (2012) Graph pattern matching revised for social network analysis. In: Graph pattern matching revised for social network analysis. ACM, Berlin/New York, pp 8–21, 26–30 Mar
18. Zeng Z, Wang J, Zhou L, Kaypis G (2007) Out-of-core coherent closed quasi-clique mining from large dense graph databases. ACM Trans Database Syst 32(2), 13
19. Dhillon IS, Guan Y, Kulis B (2007) Weighted graph cuts without eigenvectors a multilevel approach. IEEE Trans Pattern Anal Mach Intell 29(11):1944–1957
20. DB-Engines Ranking – popularity ranking of graph DBMS (2015) DB-Engines Ranking – popularity ranking of graph DBMS. http://db-engines.com/en/ranking/graph+dbms. Accessed 16 Jan 2015
21. Customers – Neo4j Graph Database (2015) Customers – Neo4j Graph Database. http://neo4j.com/customers/. Accessed 16 Jan 2015
22. Stolerick K, Florida R (2006) Creativity, connections and innovation: a study of linkages in the Montreal region. Environ Plan 38:1799–1817
23. Romano NC Jr, Nunamaker JF Jr (2001) Meeting analysis: findings from research and practice. In: Proceedings of the 34th annual Hawaii international conference on system sciences. IEEE. Maui, Hawaii, USA

Index

© Springer International Publishing Switzerland 2015
M. Trovati et al. (eds.), *Big-Data Analytics and Cloud Computing*,
DOI 10.1007/978-3-319-25313-8

Printed in the United States
By Bookmasters